Growing and Propagating Showy Native Woody Plants

Richard E. Bir

Growing and Propagating Showy Native Woody Plants

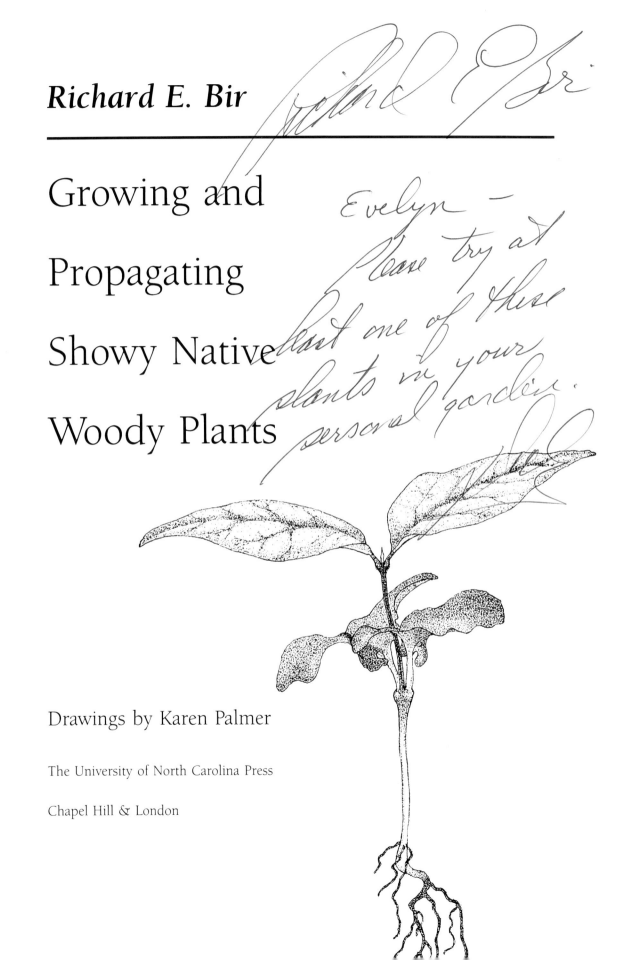

Drawings by Karen Palmer

The University of North Carolina Press

Chapel Hill & London

© 1992 The University of North Carolina Press
All rights reserved
Manufactured in Hong Kong

The paper in this book meets the guidelines for permanence and durability of the Committee on Production Guidelines for Book Longevity of the Council on Library Resources.

96   95   94   93   92   5   4   3   2   1

Library of Congress Cataloging-in-Publication Data

Bir, R. E.
  Growing and propagating showy native woody plants / by Richard E. Bir : drawings by Karen Palmer.
        p.   cm.
  Includes bibliographical references and index.
  ISBN 0-8078-2027-X (cloth : alk. paper). —
ISBN 0-8078-4366-0 (pbk. : alk. paper)
  1. Native plant gardening. 2. Native plants for cultivation—Propagation. 3. Ornamental woody plants. 4. Ornamental woody plants—Propagation. 5. Landscape gardening. I. Title.
SB439.B49   1992
635.9′5175—dc20   91-35993
                                         CIP

To the plantspeople who have taught me and to the North Carolina Cooperative Extension Service for putting me where I can share what I was taught

Contents

Preface   vii

Introduction   1

Part One. Natives in the Garden   7

1. Getting Natives Started   13
Propagation   14
   Seeds   14
   Cuttings   29
Pruning & Transplanting   43

2. The Landscape   45
Site Analysis   46
Soil Preparation   47
   Soil Physics   47
   Soil Chemistry   49

Planting   52
Mulch   56
Competition   59
Fertilization   63
Pruning   66

Part Two. Plant List   69

Appendixes

1. Landscape Fertilizer Rates   161
2. Landscape Sizes and Germination Requirements   163
3. Showy Native Shrubs and Trees for Moist Sites   169
4. Showy Native Shrubs and Trees for Dry Sites   171
5. Showy Native Shrubs and Trees That Attract Wildlife   173
6. Showy Native Shrubs and Trees That Will Tolerate Neutral or Slightly Alkaline Soils   175
7. Sources of Rooting Hormones and Other Horticultural Supplies   177
8. Sources of Seeds for Native Woody Plants   179
9. Nursery Sources of Native Woody Plants   181

Selected References   183

Index of Scientific Names   185

Index of Common Names   187

General Index   191

Preface

How does a book like this come into being?

Before it can begin to take form, at least two parts of the puzzle must exist. First, there must be need. If the subject has already been adequately covered by someone else or is a subject that no one finds interesting, then no need exists. This doesn't stop everyone, but it would have stopped me. Second, there must be someone who is willing to go to the effort required to build the book. Writing, I find, is an important but relatively small part of book construction. Once the need and the author exist, a catalyst must set the process in motion and a supporting cast must bring it to fruition.

I have been learning about horticulture and attempting to share the knowledge I've gained for most of my life. Since 1972, I have been employed doing just this. It's great work if you are lucky enough to find it.

In the process of learning about plants, I have found it impossible to ignore the fact that many of the most terrific plants are native to eastern North America. As I share this knowledge with professional horticulturists—whether one-on-one in a parking lot or as an imported speaker in front of hundreds in an ego-enhancing spotlight, using electronic voice amplification—the same questions repeatedly arise. One of them has been, "Which one book would you suggest for the gardener, the beginner?"

Quickly, I can rattle off eight or ten books that are "essential." However, I could never name just one, and none of the ones I might rattle off is really for folks who are not already formally trained or conversant in the languages of botany and horticulture. After this book is published, I still will not be able to name just one text that can answer all questions, but I hope this book will be a starting place for coming to know the joy of working with native plants and appreciating those plants in particular and horticulture in general.

A few years ago, I received a call from David Perry of the University of North Carolina Press. He had been talking with some of my friends in botany and horticulture, searching for someone to write a book on gardening with native woody plants. I was asked to suggest someone.

Eager to help, as well as to be able to have a handy reference, I started sending out letters and making phone calls. Most of them came back saying, "Dick, you are the one to write that book."

My excuse for not writing a book, up to this point, had been that the nature of my university appointment did not allow me to write a book such as the one David was after. But when Ken Moore and J. C. Raulston, both friends employed in my university system, encouraged me to write the book and disagreed with my understanding that policy restricted me from doing so, I wrote to my bosses. They took away my excuse.

I thought a book of the kind envisioned should contain black-and-white drawings as well as color photographs. Friends who had seen my photography convinced me that I could get good enough shots to take care of the latter sort of illustrations. That left the drawings. Finding an artist in the mountains of North Carolina is not a problem. Finding the right artist was something I didn't know how to do, so I turned to friends who know about such things. Karen Palmer's name surfaced a couple of times, so I went to sip tea with my former neighbor Karen. After talking around the subject of who might be a good artist to approach, Karen said, "I'd like to give it a try."

I called David. This book is the result.

Everything I do has an enormous supporting cast. I am a case study in interdependence. To single out folks in all the places where I have learned, including all the plantspeople, regardless of what they call themselves, puts me in too great a danger of forgetting someone I meant to thank. Therefore, thanks to all of you.

The list of people, many of whom I do not know, who have provided me with a wonderful classroom as well as an outdoor photographic studio is extensive. Thanks to those at the Blue Ridge Parkway, Pisgah and Nantahala National Forests, Biltmore House and Gardens, Botanical Garden at Asheville, North Carolina Botanical Garden, North Carolina State University Arboretum, Mountain Horticultural Crops Research Station, State Department of Transportation, State Division of Highways, and all the private gardens for your hard work. The place looks great!

The list of people who should be thanked in print as well as personally must start with my family: those who encouraged a love of nature in an overgrown kid and those who have tolerated being dragged through yet another garden or having the car stopped on the way to somewhere important in order to take a picture of some roadside flower. The foundation certainly was laid by Bill and Georgene Bramlage and by a plant ecology class taken before most folks had ever heard the word "ecology." Time spent as a student at Longwood Gardens and living in Great Dismal Swamp set a direction. Gil Whitton together with the Institute of Food and Agricultural Sciences at the University of Florida and later Gus Dehertogh together with the North Carolina Agricultural Extension Service took a chance on a guy whose credentials might not have exactly matched the job requirements. Thanks.

The group I've saved for last includes all those who have taught and tolerated me in the mountains of North Carolina and at North Carolina State University. Jim Shelton, Harry Silver, and Ken Perry laid out the path before I ever arrived. All I've tried to do is follow it and make it a little clearer. Ann Klimstra and Joe Conner have made it possible for me to follow the path as I have. Thanks to each of you. This has been a long-term cooperative effort.

Growing and Propagating Showy Native Woody Plants

Introduction

During my years in public horticulture I have been asked over and over, "What is that flowering shrub or tree?" or "Are these berries good to eat?" Most often, the question following these two is, "Can I grow it in my garden?"

When I answer, "Yes, that plant is native to your area," questions follow in rapid succession: "How do you grow it? Where can I get one? Will it attract butterflies? songbirds? wildlife?"

In writing this book, I hope to help nonbotanists or nonhorticulturists recognize many of these eye-catching native plants. I can't possibly include all of the native flowering and fruiting shrubs and trees that grow between the surf and the mountain peaks of the Southeast and mid-Atlantic states. There are too many. The ones I have chosen are those that I think most likely to arouse your curiosity when you see them in the wild as well as a couple that have exceptional potential for use in landscapes even though you are not likely to see them growing wild.

The land from the Atlantic shores to the crest of the Appalachians has been blessed with more native plant diversity than many continents. European explorers and plant breeders worldwide have recognized the beauty and utility of our native plants, collecting them for their pleasure as well as for research gardens. In fact, some European countries used to consider these plants great treasures not to be shared with the enemies of the realm. Now, a new generation of explorers is rediscovering the natural wonders of the region. This book is intended to help you increase your pleasure as you explore the region and learn about some of our native plants.

For over a decade I have been fortunate to work full-time with nursery growers and some of the finest plant researchers in the world. We have been trying to unlock the secrets of propagating and successfully growing those native plants with the greatest garden promise as well as finding better ways to grow some that have already found their way into many gardens. I've tried to distill much of the knowledge that I've gleaned from soil chemists in the laboratory and wildflower enthusiasts in the garden, both of whom love the land and our wonderful native plants. This continually evolving knowledge is what I'll refer to when telling you, mostly in nontechnical terms, how to grow these plants. Remember, though, there are lots of correct ways to grow things. The plants in your garden have

Map 1. Native Range of Plants Discussed

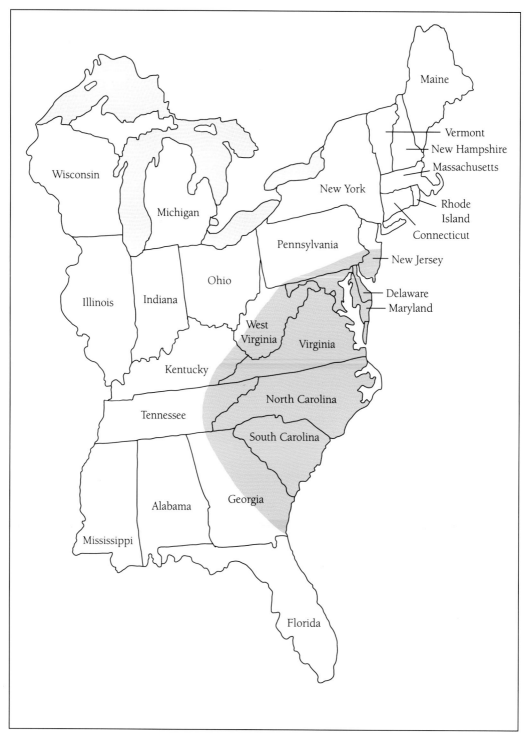

not read this book, so they shouldn't be held responsible if they don't follow my rules.

This book reflects my own strong opinions about which native plants belong in the average home landscape and which do not. Black locust, *Robinia pseudoacacia*, does not. If you want to plant one of these thorny, invasive beasts in your yard, be my guest. Leaves will drop in August; branches will seem to drop with every breeze; thorny root sprouts will even grow up through patio bricks. On the other hand, it will provide you with pretty, fragrant flowers in the spring, some of the best-tasting honey anywhere, and, eventually, some nearly indestructible fence posts. In contrast to black locust, our native dogwoods and redbuds, in addition to rarer gems like silverbells and sourwood, are small trees that belong in any North American landscape.

Determining the name of a plant that you want to know more about may be the most difficult part of using this book. Once you know the names of some native plants, figuring out which one you saw becomes easier. However, you have to know a holly from an azalea before you can tell which holly or which azalea you are looking at. Without learning much botany, perhaps the best thing to do, however time-consuming, is thumb through the book looking at the pictures. If the flower in the picture is pink but you saw a white one just like it, read the text. Many of our pink-flowering azaleas and rhododendrons have white forms.

Fortunately, most woody plants appear in the wild only under certain circumstances. Therefore, one of the first questions to ask when trying to identify a plant is, "Where did I see the plant?" Was it in the mountains, piedmont, or coastal plain? Was it a wet or dry site? Was it in the shade or in the sun?

Another fortunate occurrence is the sequence of bloom. There is a natural order to things. Dogwoods and silverbells may overlap in bloom, but both finish blooming before mountain laurels start. Therefore, it is important to get in the habit of asking both when something was flowering and what else was flowering at the same time.

Spring is supposed to move north at the rate of fifteen miles per day, and up mountains at a rate of 1 day for every 100 feet in elevation. To complicate this, we're told that for every 1,000-foot increase in elevation the temperature decreases by 3.5°.* However, if you see a plant on the north side of a mountain, that 1,000-foot change—from Greenville, South Carolina, to Asheville, North Carolina, for example—is going to produce a temperature drop of closer to 10°. As a result, a plant in Asheville will often bloom a week later than a plant of the same species in nearby Greenville. If a plant seems to be flowering earlier or later than you would expect it to on the basis of weather, look for nearby rocks that warm in the sun, large bodies of water like lakes and rivers that have a cooling effect, buildings that block the sun, or other topographic features that can influence when plants flower. Despite these local "modifiers" affecting the time at which particular species flower, the sequence of bloom among species generally remains the same.

*All temperatures given in this book are Fahrenheit.

The last things to remember when trying to identify a plant are what color the flower or fruit was and whether it had any fragrance. Knowing that a plant had red berries may not seem very important, but it at least eliminates plants with black, blue, or white berries. If you also know where and when you saw the plant, and how big it was, you only have to look at a few pictures before you can be pretty sure of what you saw.

The photographs in this book were taken with natural light, outdoors, where the plants were growing. This means that they feature shadows, leaf spots, and an occasional insect. Because no reflectors or artificial lights were used, the photographs more accurately reflect what is actually seen along the roadside.

I want to share at least one successful way to grow and propagate each of the native plants described in this book without making a major investment in the expensive supplies and equipment used by professional horticulturists. When a spot in the landscape cries out for change, I usually want only a couple of plants, not thousands. Most of us have room to propagate a couple of plants.

If you want to grow thousands of plants, you may want to get into the nursery business. If so, contact your county agricultural extension agent. The plants may remain the same as in gardening, but the techniques used for growing most of them as well as the level of anxiety involved will change.

You will need to learn a few common horticultural terms in order to understand some parts of this book. For example, if you want to propagate redbud (*Cercis canadensis*), stratifying the seeds isn't enough. To get them to sprout and grow you must also scarify the seeds. You need to know what both scarify and stratify mean if you expect to be successful with your redbud seeds. Both terms are explained in the seed propagation section. If you encounter other terms that are unfamiliar as you read the text, refer to the general index to locate explanatory discussions of these terms.

Many gardeners who choose native plants wish to grow them without using chemical fertilizers and pesticides. This book is not the place for an argument about which techniques are natural or organic and which are not. However, I will offer alternatives to standard production techniques for those who want to avoid man-made chemicals.

Most of us inherit the big trees and many of the shrubs in our landscapes. We personalize our yards just as we do the insides of our homes by adding both beauty and utility. The shrubs and trees that have flowers or fruit are the ones that attract us most and that attract the wildlife we find so interesting (who hasn't enjoyed the antics of squirrels and songbirds outside the window?).

The plants I have included are showy, that is, something about them attracts attention. The prominent feature may be flowers, fruit, colorful foliage, fragrance, or even fluff clinging to seeds. These interesting, seasonally showy plants seem to be in scale with our surroundings and in keeping with our lifestyles, adapting to formal or more relaxed landscapes. Most are small enough to leave us room to live but substantial enough to provide the landscape framework that enhances our homes. They provide protection from noise, dust, wind, and the hot sun, while entertaining us with their pleasing forms or inviting other performers to our stage.

Plants don't respect political boundaries and often find ways around natural ones. Frequently a river or mountain divides governments, but the same species of plants grows on both banks of the river and both sides of the mountain.

Although the plants that are the subject of this book are native to the area of the eastern United States shown in Map 1, they can be found growing wild from Florida north into Canada. They are most common, of course, where conditions resemble those under which they are found in greatest abundance, in similar habitats and hardiness zones in states bordering their native region. The natural diversity of vegetation in this region encompasses everything from coastal palmettos to fir and spruce forests like those in Canada. Many of these species and their descendants are valued in Asian and European gardens as well as throughout the Americas. Native plants from our region have influenced gardening on a truly global scale. We should appreciate these plants at home as well.

Part One

Natives in the Garden

Native plants belong in your garden for many reasons. The best reason is that they are good plants for the landscape. Many natives, like many exotic species, will tolerate a wide variety of landscape conditions, providing the flexibility necessary for successful home landscapes. My yard has wet and dry spots, hot and cold areas. There is never enough space, so plants get crammed together in my "American cottage garden," competing with each other and contending with the activities of a family, neighborhood pets and wildlife, and soil that is never quite what it should be.

Even though some plants' natural habitats are very restricted, when brought into the garden and given the minimal care that most of us provide for the rest of the landscape, many native plants will thrive or at least hold their own. A good example is the pinkshell azalea, *Rhododendron vaseyi*. It is native to only a few counties of the North Carolina mountains and is rarely found growing wild below 3,000 feet in elevation. How could something so "rare" be worthy of a place in your garden?

First of all, many uncommon plants in the wild are not "rare." On a May drive along the Blue Ridge Parkway and down Highway 215 in Transylvania County, North Carolina, I saw at least 5,000 pinkshell azalea bushes in bloom over a stretch of about twenty miles without ever leaving the road. Nature provided a show to take the breath away. Somehow "rare" doesn't seem like the right word for a plant that dominated the flowering plant display over many miles that day. Perhaps because of where this azalea is native (i.e., the exposed cliffs and oak-shaded forests of the southern Blue Ridge), it has been an excellent candidate for moving to other, much colder climates.

Hardiness zones are one of the ways horticulturists describe how much cold a plant can tolerate. They are determined by the average minimum temperature in an area. Each 10° drop in temperature places a region in the next-lower-numbered hardiness zone (Zone 11 being that with the highest average minimum temperature and Zone 1 that with the lowest). Asheville, North Carolina, and Roanoke, Virginia, are in Zone 6. Columbia, South Carolina, Raleigh, North Carolina, and most of Tidewater Virginia are in Zone 7. Much of coastal North and South Carolina are in Zone 8, but folks lucky enough to live along the coast near Charleston may claim Zone 9. Remember, however, that these zones are based upon averages. I've lived in the Blue Ridge for a little more than a decade, but I've never seen an average year. Map 2 shows plant hardiness zones for much of the eastern United States as determined by the United States Department of Agriculture.

It is important to remember that the map reflects average temperature extremes, not overall average temperatures. However, these extremes of cold are often what limits a plant's survival.

The map can provide insight into which plants might be suitable for your garden or that of distant friends. For example, pinkshell azalea is native to Zone 6, yet I know of pinkshell azaleas thriving in gardens in Zones 4 and 5. Perhaps this is a southern beauty to send to your friends up north. Pinkshell isn't supposed to move south well, but I've seen plants covered with flowers in the heat of Raleigh. In areas

Map 2. Plant Hardiness Zones for the Eastern United States

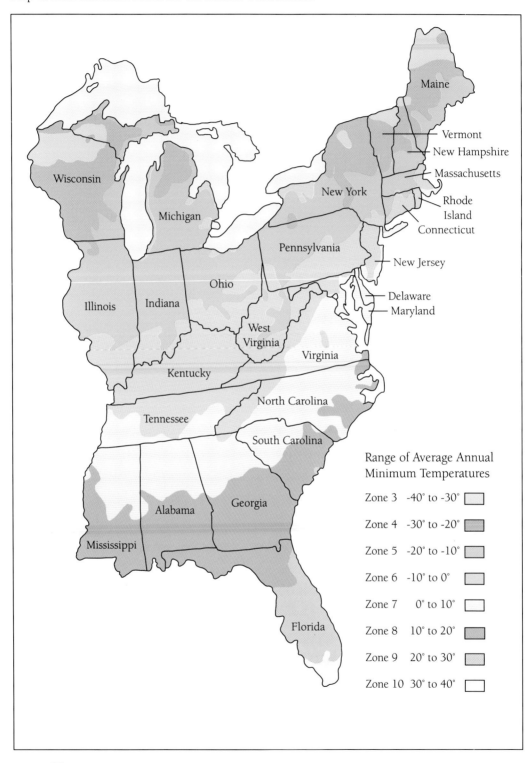

10

warmer than their native mountains, pinkshell azaleas must have shade from the afternoon sun and a good organic mulch to keep moisture even around the roots.

Azaleas have a reputation for dying if they have "wet feet," yet pinkshell often grows wild in bogs or in "seeps" filled with sphagnum moss at the bottom of mountain cliffs. However, it is one of those plants that can tolerate a wide variety of conditions. Established pinkshell azaleas survived the terrible droughts of the 1980s, when the soil turned to dry powder and the oak trees providing their shade started to die.

Why isn't pinkshell azalea more widely known? For one thing, it is a deciduous azalea, meaning it is bare of leaves in the winter. This worked against it in the 1950s and 1960s, when we were taught that good landscape shrubs must keep their leaves all year. What rubbish! Further, it blooms a week or so before flowering dogwoods. Timid souls that we are, we find it a bit too chilly to linger outdoors at that time of year. If planted where you can see it from inside your warm home looking out into a wooded setting on a gray spring day, however, pinkshell is guaranteed to buoy your spirits. Finally, pinkshell has no fragrance.

Pinkshell azaleas and many other native plants have difficulty surviving when transplanted from the wild. Your chances for success with a plant collected from the wild are limited even if you are willing to baby the plant for at least a year.

Additionally, you should not remove plants from the wild because you are stealing unless the plant was growing on your property or you have permission from the landowner and because many plants that are rare or endangered will grow only under very special conditions. Sometimes the act of removing the plant you desire is enough to destroy those conditions. Even plants that are not rare or endangered may never reestablish themselves in the area to which they are moved. The only reason for transplanting a native woody plant from the wild to your garden is in order to save it when you know that the area in which the plant is growing is about to be destroyed. Home, business, and road construction probably threatens more plants via habitat destruction than plant collectors ever could.

The North Carolina Botanical Garden in Chapel Hill has been promoting the concept of "conservation through propagation" for years. They are asking gardeners and nurseries alike to propagate native plants, grow them to landscape size, and plant them. All that is ever taken from the wild is a few seeds or an occasional cutting from particularly desirable plants. Plant populations in the wild are not endangered by this practice. For gardeners, however, there is a better reason to practice conservation through propagation. Nursery-grown plants are better plants. They have been grown under conditions that are as good as the nursery workers are able to provide. They have healthy leaves and roots. Usually they have been pruned and fed, so they look like good landscape plants as well as have an excellent chance of surviving in your garden with only the minimal care needed to establish any plant worth having.

The end result of the North Carolina Botanical Garden program is that there are more rather than fewer native plants. Nursery or home propagation helps to ensure that native stands of plants remain for everyone to enjoy, preventing erosion and feeding wildlife, while the same species grow in our gardens, feeding our senses and attracting butterflies and bumblebees. Pinkshell azalea is just one of the native plants that has been adopted by a few nurseries. If you can't find the woody native plant you want locally or in a mail-order catalog, ask your landscaper or garden center that normally carries native plants. It may take a while, but they should be able to find one for you.

Woody plants are the shrubs and trees that live for years. Rather than dying to the ground each year like herbaceous perennials, they have stems made of wood. All too often, good information about woody plants for home gardeners is either hard to find or difficult to understand without a degree in botany or horticulture. This is unfortunate (and a situation I hope to remedy by telling you about these plants in language you can understand) since shrubs and trees provide the backbone for most landscapes. They represent an important and often sizable investment. Properly selected native woody plants are well adapted to our region. As a result, they will succeed in your garden.

One

Getting Natives Started

Gardeners who plant a tree or shrub seed and then end up with a good landscape plant where they want it, in their lifetime, are the rarest and luckiest of people. The conditions needed for a young plant to thrive aren't necessarily the ones that are best for the plant when it reaches flowering or fruiting age. Competition in the landscape or the care we provide when growing plants in a crowded, amended garden can often stunt or kill young plants. We wouldn't think of sending newborn babies off to school, but expect our older children to grow and thrive when they go to school. Plants, like people, having changing needs and react to changes in their environment.

Nurseries are just as important for the growth of woody, perennial plants as they are for other babies. Plants require much more care when tiny than when mature. In fact, professional nurseries are often divided into propagators, transplant (liner) producers, growers of garden-center-size (easily portable) plants, and growers of large landscape specimens. In the section on propagation later in this chapter, I'll try to give you enough information to get plants up to the size that you can put them into the landscape without the help of expensive earth- and plant-moving equipment.

Perhaps the most important thing to keep in mind when introducing a native plant to your landscape is a vision of the plant in the wild. In both propagating and later growing plants in the garden, we are trying to provide conditions that permit them to grow and thrive. We do not "make" a plant grow; plants grow in response to the conditions we provide. Therefore, it is important to pay attention to where little plants grow as well as to the beautiful, established plants that inspire us. (Coveting is a normal part of gardening. If something is beautiful somewhere else, it is perfectly all right to want to have that beauty where you can enjoy it regularly.)

Please remember that woody plants are perennials, meaning that they live for years. With proper care, many will outlive us. As with many other things of value that last a long time, patience is needed to bring a woody plant to its fullest value in the landscape. Annuals such as marigolds and zinnias will grow, flower, set seeds, and die in the space of a few weeks or months. This is their natural cycle. In that same time, we may not even have satisfied the conditions needed for the seeds of trees such as Carolina silverbell to sprout.

Propagation

Plant propagation is simply what you need to do to make more plants. Many techniques exist. Some methods require conditions and equipment that would be the envy of a hospital laboratory. The only two methods I will discuss in any detail are sexual (growing plants from seeds) and asexual (rooting cuttings) propagation. I'll say more about the requirements of specific plants in Part 2 of the book. In this section I want to be sure we are all starting with the same basic knowledge.

Why bother to propagate plants at all? Why build furniture? You can buy furniture. Why play tennis when you can watch others? For some, propagation is interesting and enjoyable, as well as a form of self-expression. Propagation may also be the only way you can obtain certain plants that aren't available in nurseries. Whether you propagate your own plants for the fun of it or out of necessity, you will develop skills, increase your knowledge, and derive a sense of fulfillment. I hope you will share the excitement I still feel when a seed sprouts or a shoot strikes roots. However, if you only want one or two easily obtained plants and you want them immediately, propagation probably isn't the answer for you. Most woody plants grow slowly when compared to what's in your lawn or vegetable garden. The nursery is one of the few places where you can buy time; a shrub or tree purchased there, ready to flower, is usually at least a few years old.

Seeds

Occasionally native plants originate when a piece of a plant breaks off, falls into the mud, and roots; but nearly all other native plants in the wild start from seeds. I find this knowledge comforting, although somewhat frustrating. It's comforting to know that the plant you are working with can be grown from seed. It's frustrating when you cannot unlock the mysteries of germination, thereby providing conditions that will cause another desirable plant to grow from seed.

All plants require certain conditions in order to grow. In the case of woody plants, some of those conditions must be met before seeds will sprout and grow (germinate). Woody plants have evolved fascinating survival mechanisms to the point that some seeds must be exposed to fire or flood or must be eaten before they will grow.

For some plants—native azaleas, for example—all you need to do to induce germination is provide the right temperature, light, and moisture conditions—in the case of azaleas, 75°, light, and damp peat. Within two or three weeks, if the seed was alive, you should see the tiny green hairs that will develop into beautiful flowering plants. For other plants, including many discussed in this book, additional inducements or conditions are necessary, and it is important to know something about the methods of scarification and stratification used to encourage germination.

Scarification. Some seeds have very thick coats that prevent moisture from penetrating

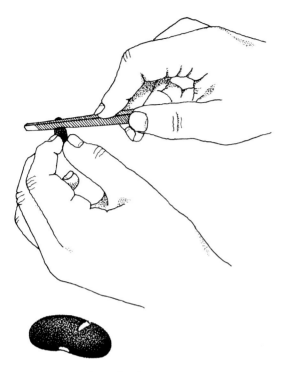

Figure 1. Seed scarification.

them. In nature, these seed coats are gradually worn away or the seeds are cracked by freezing. Once the integrity of the coating is broken, moisture can penetrate the seed, allowing the plant to grow. In a man-made environment, scarification simulates the natural breaking down of the seed coat by artificial means.

Fortunately, there are many ways to scarify seeds. The simplest is to file or cut a notch in the seed coat, barely penetrating the outside of the seed. Look for an eye (hilum), which is the scar from where a seed was attached to the mother plant, and then penetrate the seed coat on the side opposite the eye (see Figure 1). When you can see a color difference, you have penetrated far enough. Although this process is slow and tedious, it is the safest for the seeds. If you use a triangular file, it is usually safest for the gardener as well. Knives sharp enough to penetrate the seed coat are also sharp enough to cut a gardener, so be careful.

If you have lots of seeds to scarify, manual scarification is not very practical. The next-safest method is to scarify the seeds with hot water. Bring water just to a boil. Remove it from the heat and wait for the bubbling to stop. Pour the seeds into the hot water or pour the water over the seeds, as long as they are in a temperature-resistant container (glass jars tend to crack, releasing very hot water onto the table and often your lap). When the water has cooled to room temperature, the seeds are scarified. In experiments using this technique compared with manual scarification, about one-third less germination occurred with the hot water technique. But then 60 percent germination with hot water is better than the less than 5 percent germination that resulted with no scarification. Even though hot water scarification produces less than the 90 percent germination achieved with manual scarification, I generally have more seeds than time. In addition, my big fingers aren't nimble enough to hold a tiny seed in one hand and file a notch with the other. If you remember that the root of the word "scarification" is "scar" and picture my hands trying to notch a tiny seed, I doubt you will confuse scarification and stratification.

When scarifying seeds, we wear away or break the seed coat. Commercially, this is often done with concentrated acids. *Do not acid scarify seeds without proper laboratory safety equipment.* The acid eats away the seed coat, generating heat and noxious gases.

The trick to acid scarification is to keep the process cool enough not to cook and kill the seeds while stopping the process just short of actually penetrating the resistant seed coat. The acid will very quickly eat away the living parts of the seed inside the seed coat if the reaction is not stopped in time. All of this tricky timing and temperature control must be accomplished while not inhaling the acid fumes. Please don't try acid scarification.

Stratification. Originally, stratification meant providing a cool period to satisfy seed germination requirements. Layers of seeds and propagation media (with very specific suggestions concerning where strata of sand or leaf mold should be placed to keep seeds moist and to drain water away from them) were arranged carefully and then cooled. These strata of media and seeds, a gardener's club sandwich, lent the term "stratification" to the procedure.

With advances in our knowledge of stratification requirements and in the development of plastics and artificial growing media, simpler techniques have evolved and been discussed in many texts. These techniques are for warm as well as cool stratification. In this book, I will refer to both kinds of stratification.

Stratification may be needed to allow a seed to sprout. Some seeds are not mature when they leave their parent plant. Often the embryo is not fully developed, so the seed needs to keep growing and developing. This can happen during warm stratification. Even though no changes are apparent on the outside, on the inside the seed is continuing to change. When embryos are fully mature, chemical changes may need to take place within some seeds before they can germinate. If all seeds came up at once, a calamity like a freeze, flood, or drought could wipe out a species. For seeds that require stratification, immature embryos and germination-inhibiting chemicals are no less important as survival characteristics than impervious seed coats are for seeds that must undergo scarification.

For stratification to work, a seed must be moist inside. If a seed requiring stratification is dry, place it in warm (70–90°) water and allow it to soak overnight. If the seed also needs scarification, scarification must take place before stratification or the water will not be able to penetrate the seed coat. Cool, dry seed storage may extend the life of a seed but will not satisfy the stratification requirements of that seed. Moisture is essential.

If you have only a few seeds that need stratification, you may want to plant them in a container before beginning the process. If warm stratification is required, as with fringe tree (*Chionanthus virginicus*), placing the container where it will not dry out but will be exposed to normal growing temperatures for the required period of time is simple enough and works. Seeds will *not* warm stratify faster if you raise temperatures above 80°, and they can be damaged by temperatures much over 100°. If you have a lot of seeds, you may want to mix them with sand or peat, making sure the sand or peat is uniformly moist but not sopping wet, and then place the mixture of media and seeds in a plastic food storage bag. You should see *no free water* in the bag. One way to reduce the chances of excess water accumulating in your plastic stratification bag is to poke a hole in the bottom of the bag with an ice pick. Free water will drain through the hole.

While working with large numbers of seeds is easier in bags than in pots, problems exist with this technique. First of all, plastic bags must be kept out of the sun. In the sun, clear plastic bags can become solar ovens that will cook and kill your seeds. Additionally, since water will not leak out of bags unless they have been punctured, it is easy to keep conditions too wet and thus cause seeds to rot. This is why there should be no free water in the bag at any time during the stratification process. I prefer to stratify most woody plant seeds with coarse builder's sand, using one measure of seeds to at least two measures of sand. When the stratification process is finished, the sand can be washed off the seeds and through a screen. Utilizing a screen while washing makes it easier to recover the seeds for sowing.

Sand can be contaminated with organisms that will cause seeds to rot, so I usually pasteurize the sand before mixing it with the seeds. This is done by pouring boiling water over the sand and letting the water drain through. Since the sand needs to be moistened anyway, the only change in procedure is to boil the water. These rot organisms may also be present in the seeds themselves, so pasteurizing the sand may or may not prevent rotting. However, I've found it worth the time and effort.

The last thing to remember about warm stratification is where you put the bag. Nothing lasts forever. I have discovered forgotten plastic bags of seeds undergoing warm stratification in a drawer years after initiating the process. Mark the day seeds are to go from warm to cool or come out of cool stratification on a calendar you see frequently so you don't forget about them.

Similar procedures are used for cool stratification and for warm. The only difference is the temperature. If your seeds are in a pot, I suggest you put the pot in a plastic bag so it won't dry out during cool stratification. Please note that *cool*, not cold, temperatures are necessary. Seeds must be at temperatures above freezing for cool stratification to work. I usually use a refrigerator set at 40°, although temperatures between 36° and 45° will work equally well for most plants. If you are stratifying in the family refrigerator, seeds should not be near fruits. Fruit can give off a colorless and odorless gas called ethylene, which is a plant growth hormone. Exposure to ethylene can prevent the germination of seeds or cause distorted plant growth.

Timing, though it can't always be precise, is extremely important. Once the stratification requirements of seeds have been met, they will sprout, even in a cool refrigerator where you don't want them to. Timing is also crucial when working with plants that have different stratification requirements for different parts of the seed. For example, after you have properly warm stratified fringe tree seeds, the part of the seed that will develop into the root, the radical, will emerge. The part of the seed that will develop into the top, however, needs cool stratification before the seeds will germinate. Therefore, after you warm stratify a bag of fringe tree seeds you are dealing with what looks like living commas. These are very fragile sprouting seeds. They must be handled gently or they can be broken. This is one reason why nurseries usually do not stratify fringe tree seeds in a plastic bag. These seeds are ordinarily planted in pots or outdoor beds before stratification begins.

Figures given here and in other books to indicate the stratification requirements for certain seeds should serve as a guide only. Each seed may have a slightly different genetic makeup. Seeds develop and ripen under different conditions every year. We know that seeds from the northern end of a plant's natural range may have different stratification requirements from seeds from the southern end. They evolve this way naturally. Discussions of provenance—beyond the simple advice that you can usually grow better plants from locally obtained seeds—are more than I can tackle in this book. So remember that when you read that flowering dogwood (*Cornus florida*) needs 60 days of cool stratification, you can count on most of the seeds germinating after 60 days' stratification at 40° most of the time. However, it's possible that you have dogwood seeds that need 75 days of stratification at 40°. It may also be that your refrigerator runs at 35° instead of 40°. In that case, you may need 90 days to accomplish what could be done in 60 days at 40°. Biological sciences are sometimes called imprecise because we cannot control all of these preconditions or predict all possible outcomes.

The way many nurseries get around these imprecise and often complicated requirements for plant germination is to use the great equalizer, nature. If you plant a viable seed and it doesn't get eaten or rot, in the cycle of seasons it will eventually be exposed to the right conditions and grow.

Seeds of plants with double dormancy requirements, like fringe tree, can be collected in the fall, cleaned of pulp, and either planted immediately or stored until spring. In either case, they will receive their warm stratification the following summer, and radicals will sprout. The next fall, winter, and spring, their cool stratification requirements will be met. When the soil warms during the second spring following planting, little fringe trees should germinate. For this to happen successfully, the area where you planted the fringe trees must be remembered and cared for, even though it will remain bare for months. You also need to know what fringe tree seedlings look like so you don't pull them up along with weed seedlings once they sprout. This natural technique takes longer but is much simpler than artificial stratification if you have the space and time. Nurseries usually have areas for germinating doubly dormant seeds covered with screens or something similar to ward off wildlife as well as mark an area where something important if not apparent is taking place.

Collection. Seeds to be used in propagation should be harvested when mature. Most seeds are mature enough to be harvested *before* they fall off the plant.

Changes occur when seeds are mature enough to be harvested. Fruits change color and soften around some seeds. The capsule or husk around other seeds may start to split. When you see these signs, harvest seeds or you may lose your chance. Blueberries, cherries, and most other fruits are favorite foods for birds, bears, opossums, and other wildlife. While you wait for your seed crop to ripen, it may be eaten. Small seeds such as those of native azaleas may be blown away as soon as the seed capsule splits open. Witch hazel (*Hamamelis virginiana*) has a particularly bizarre seed dispersal mechanism, which you need to know about in case you are interested in trying to propagate this plant. The plant flowers in fall or winter, but seeds are not dispersed for a full year. About the time witch hazels are getting ready to flower, the previous year's seed crop comes flying out of the seed capsule, expelled with such force that if one hits you, you may think you've been hit by a small stone or flying insect.

Seeds, except those in mushy fruit, should be collected in a paper sack rather than a plastic bag. Once seeds have been collected, you want to keep them alive. Storing seeds where they will get too hot (such as in a tightly closed car parked in the sun) is one excellent way to kill them. Keeping seeds wet so they rot is another. Seeds in closed plastic bags cannot dry.

If seed capsules have not opened, letting them dry in a warm (up to 105°) place will sometimes cause pods to split. I usually keep them in a paper bag while they're drying unless I have large amounts of seed or seed capsules and pods are too wet from having been collected in the rain or a heavy dew. If I have too many seeds for them to dry thoroughly in one crowded paper bag, I ordinarily transfer them to lots of paper bags when I get home from collecting. If seeds or pods are wet, I collect them in plastic and then try to spread them out to dry soon after getting home. When their surfaces are dry, I usually put them in a paper bag for further drying. Once pods or capsules are dry, shaking them in a paper bag will help the naked seeds fall out of the pod. If this does not work, it may be necessary to crush seed capsules to extract the seeds. When removed, naked seeds should be spread out on paper so they can dry in a warm (under 105°), not hot, area.

Seeds harvested inside fruit should be removed from the fruit as soon as possible after harvest. Fruit may include natural chemical germination inhibitors, which, if allowed to dry on the fruit, will delay, reduce, or prevent seed germination.

Techniques for removing seeds from fruit vary. Start by trying to squeeze the seeds out as soon as the fruit is soft enough to crush. If this cannot be done easily, soak the fruit in water that is at room temperature (70–90°) for 24 to 48 hours to soften it. Because of the odor generated by fermentation, the water-fruit mixture should be kept outdoors, preferably in the shade, and covered to protect it from insects and foraging wildlife. I like to use fabric or netting to cover the jar, using an elastic band to hold it in place. The fabric allows the mixture to "breathe." Tightly closed jars containing a fermenting solution have been known to explode.

If no bubbling shows 24 hours after mixing the fruit and water, add a packet of yeast to encourage fruit fermentation and decomposition. Stir the mixture while the brew is "working" to separate the fermented and decomposing fruit from the seeds. After the brew has fermented 24 to 48 hours, stir it and pour off floating pulp and seeds. The good seeds should sink to the bottom while hollow seeds float on the surface (see Figure 2). Small seeds may have to be separated manually. Any remaining pulp can usually be removed from good seeds by rubbing the seeds on a screen under a stream of water. Once the seeds are reasonably clean, spread them in the shade to dry like naked seeds. Don't be too rough. Your seeds are alive. Rough handling may damage some vital internal part, thus preventing germination.

Storage. The longer seeds of most native woody plants are stored, the lower the percentage of germination will be. Therefore, measures to keep seeds fresh and viable are important.

Once most woody plant seeds have been cleaned and the surface of their seed coats dried, they can be safely stored at room temperature for a few weeks. If you need to store seeds longer, reducing the temperature will extend seed life. Freshly harvested seeds contain all the food this living part of the plant will have until germination is complete and new green leaves can manufacture more food for the plant. Anything you can do that helps to maintain these food resources until they are needed for seed germination will increase your chances for successful propagation.

If seeds will be used the season following collection, storage in a moisture-proof container at cool temperatures above freezing is acceptable. Cool temperatures slow down seed metabolism, extending the life of the seed. Larger seeds usually have a shorter storage life than small seeds. I use glass jars to store seeds whenever possible. Glass is easy to clean and disinfect.

Longer-term storage should also take place in moisture-proof containers. Sub-freezing temperatures should be used for the seeds of species that won't be hurt by such cold storage. The 0° temperature of most freezers is excellent for long-term storage of these seeds. Refrigerated storage (33–40°) is safe for seeds of all plants mentioned in this book, but do not store seeds along with ethylene-producing fruit such as apples. Seeds won't hurt the fruit, but ethylene can hurt your woody native plant seeds.

Sowing and Growing. If seeds of a particular plant have special germination requirements like light, dark, scarification, or stratification, those requirements are noted in the plant list or in Appendix 2. Otherwise, normal springtime gardening conditions should be adequate for good germination.

I suggest sowing all small and tiny seeds in containers on artificial growing media

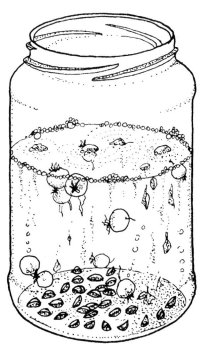

Figure 2. Separating seeds from fruit by fermentation.

composed of whatever combination of bark, peat, perlite, and vermiculite suits you. I generally use three parts (a bucketful usually equals one part when I'm mixing growing media) southern pine bark that has been passed through a ¼-inch-mesh screen to one part sphagnum peat. I use the cheapest grade sold as pine bark mulch and mix it thoroughly with the peat. To my 3:1 bark-peat mixture I add about ¾ pound (one slightly rounded cup) of dolomitic limestone per 3 cubic feet. Since the bark comes in 3-cubic-foot bags, the bags are a convenient measuring device. I just refill the bag with the mixture to be sure I have the right amount of media, add the dolomitic lime, and then thoroughly combine the lime and media, usually on the garage floor with a clean spade. If you have a large space like a garden cart or wheelbarrow that can be cleaned before mixing, it will work just as well.

Unless you are going to be growing large numbers of native seedlings, mixing your own potting soil is probably not worth the trouble. I suggest you buy a small bag of commercial potting mix suitable for azaleas instead. The important point to remember is that your media should not contain "real" soil. Real soil usually does not drain well, and it can contain insects, diseases, and weed seeds. Pasteurization can correct the last three problems but will not improve poor drainage.

Fill your germination tray or pot to within 1 inch of the rim with potting media. Moisten the media. Wetting some soilless mixes may be difficult. Placing the pot in a pan or saucer and then *gently* pouring hot water over the top of hard-to-wet media before sowing the seeds may help uniformly wet the mix. Let the media sit for a couple of hours in the pan or saucer filled with the water that drained through the mix. Then discard the water and place the pot back in the pan, allowing the media to drain for another hour. Discard any water that collects in the pan.

Once the mix is moist and cool to the touch, sow your seeds and cover them to a depth four times their width. For example, a seed that is $1/16$ inch wide would be covered $4/16$, or $1/4$, inch deep. I often use pure vermiculite to cover the seeds. Very tiny seeds, like those of mountain laurel or azaleas, shouldn't be covered at all. After seeds have been uniformly sown, gently moisten the surface of the growing media and cover the pot with a plastic bag (see Figure 3). Place the plastic-covered pot in a warm, well-lit location but *not in direct sunlight*. If everything is in order, germination should begin in 1 to 3 weeks, depending upon species and temperature.

A day or two after you notice the seeds sprouting and standing up on their own, remove the plastic bag. Rather than removing the bag all at once, I usually poke or cut holes in it to let the seeds gradually get used to less humid air for a couple of days; then I remove the bag. With zip-top bags, gradually increase the size of the opening instead of puncturing the bags. Your pots should not require watering before seeds germinate, but if they do, place the pot in a bowl or pan, fill the pan with water, and then go away. Return in an hour to discard any water that remains.

After seedlings are up, move them at night to a place where they will receive direct sunlight. Moving seedlings at night allows them gradually to get used to drier air and brightness the next morning. I prefer an east- or north-facing window for the first few days. Once seedlings of sun-loving plants are adjusted to bright light, they can be placed on a south- or west-facing windowsill. If you are lucky enough to have a greenhouse, shade the seedlings for the first few days and then remove the shade cloth at dusk or during a cloudy day.

I start fertilizing seedlings when the first true leaves appear. Use a quality house plant food such as Peters, Miracle-Gro, or Rapid Grow at one-half the rate suggested on the label. If the seedlings are receiving at least 6 hours of sunlight each day, half-rate fertilizer is applied every 7 to 10 days for 6 weeks. Then full-strength fertilizer is used until the seedlings are ready to transplant. Most of the small-seeded woody plants I grow are ready to transplant 3 months after the seeds are sown. For this reason, seeds are often sown in winter so they will be ready to transplant in spring. As plants get larger, space is usually scarce, so moving plants outside as soon as the danger of frost passes is my standard procedure.

Organic growers can fertilize seedlings with manure teas made by adding water to manure and letting the mixture steep until it achieves the color of tea. Although I've used manure tea with some success, I'm not fond of its bouquet. I've also had good results with cottonseed meal and fish emulsion as fertilizers. Other organic fertilizers have yielded variable results for me, so please experiment with only a few plants before risking your whole crop to a new and unknown fertilizer. Too much organic fertilizer can kill plants just as surely as too much chemical 10-10-10.

Large-seeded plants can certainly be grown in pots or trays, but I prefer to grow them outdoors to take advantage of natural conditions. Why? It's easier, and the resulting plants are often better. Pots may require daily watering, but seedbeds rarely require

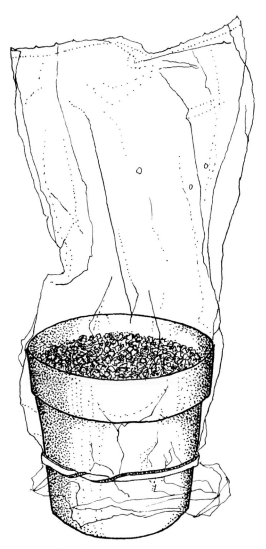

Figure 3. Plastic bag covering pot to promote seed germination.

watering more often than weekly once germination is complete. You also don't need to worry as much about light or space or about plants being stunted if they outgrow the pot before you can get around to transplanting them. The 2 square feet of garden space needed to grow 40 healthy flowering dogwood seedlings seems like nothing, but 2 square feet of windowsill space is a lot.

If you decide to grow your plants in an outdoor seedbed, choose and prepare your site carefully. Select a site with good garden soil that receives at least 6 hours of sunlight daily unless you are growing plants that normally grow in deep shade in the wild. Most shade-loving plants should be shaded throughout their youth and in warm seasons. Water should drain through the soil on your seedbed site fast enough so that no puddles remain an hour after a heavy shower. If your soil is too dense, heavy, or tight, spread pine bark mulch about 4 inches deep over the area where you will be planting seeds and then mix the pine bark thoroughly with the topsoil, using a rototiller or spading fork. Be sure to break up clumps of soil and remove rocks, large roots, and other obstacles to seed germination. After preparing the soil in this way, rake the surface of the bed smooth.

Contact your county agricultural extension agent at least a month before planting to learn how to have your soil fertility tested. Usually you will need to send a soil sample to a state or private laboratory for analysis. You will have your soil test results back from the lab in a few weeks during most of the year. I like to send samples to the lab around Labor Day so I have the results in time for fall planting. For spring planting, submit samples before Christmas because soil labs get busy during the winter testing for commercial agriculture. You may not get results from samples submitted in January until March.

When soil test results come back from the lab, spread any limestone or fertilizers that may be suggested uniformly on the surface of your plant bed. Next, turn the soil to a depth of at least 6 inches, mixing fertilizer and soil thoroughly. Rake the bed smooth and either level or slightly higher in the middle than on the edges to allow for settling soil (see Figure 4). Avoid leaving low spots, as water tends to puddle in them even in bark-amended soil.

If seeds require stratification, plant them in the fall. Nature will cool stratify them for you over the winter. If seeds have a double dormancy, the need for a warm period will be satisfied in either the fall or the following summer. Seeds needing scarification should be scarified before they are sown.

After sowing seeds, cover them four times as deep as the seed is wide with good garden soil. If your soil is a heavy clay, cover the seeds with peat, softwood (fir, pine, etc.) sawdust, or vermiculite; firm the covering over the seeds with a lawn roller or by placing a board over the sown area and then stepping on the board to uniformly firm the soil. Firming will ensure the intimate contact needed between seeds and soil. Once beds are firmed, irrigate lightly.

When the weather starts to warm in the spring, soil temperatures will also warm and seeds will sprout. Be sure that the soil stays moist (but not soggy) at this time or germinating seeds may dry out and die before penetrating the soil surface. The month when you must be particularly careful to prevent topsoil from drying just as seeds are germinating moves inland and from south to north in the spring. It's ordinarily March along the coast, April in the piedmont, and May in the mountains, with lots of variation due to site and season. Once seeds have germinated, irrigate less often but for longer periods of time. Irrigate only when plants need it. Once seedlings have two true leaves, irrigate if the soil feels dry when your index finger is pushed into it about an inch. If rainfall is regular, you may never have to irrigate after the critical spring germination period.

If seeds germinate but then seedlings flop over even though there is plenty of water in the soil, or if seedlings disappear entirely,

Figure 4. Seedbed slightly mounded in the middle.

contact your county agricultural extension agent or garden center or talk to an experienced gardener. You may be faced with a treatable disease, foraging insect, or wildlife problem, and these "experts" should be able to advise you concerning the proper course of action, if one exists.

I rarely apply much fertilizer to seedbeds of native woody plants the first season, preferring to let seedlings develop at their own pace. However, more rapid growth is sometimes desirable. As long as it's at least 6 weeks before the average first-frost date in the fall, growth can be accelerated by an application every two weeks of a balanced fertilizer applied as a liquid via sprinkling can or siphon proportioner (e.g., Hozon—see Appendix 7 for sources).

If you choose to use a balanced granular fertilizer such as 10-10-10, be sure the fertilizer is applied when the leaves and stems of plants are dry and then brushed or watered off the leaves so that the fertilizer is on the soil, not touching tender young plants. If granular fertilizer remains in contact with the plant, a "burned" or dead plant may result. A light application of 10-10-10 would be ½ cup per 100 square feet (a bed 4 feet wide by 25 feet long). (See Appendix 1 for fertilizer application rates.) After applying fertilizer, take care to water plants thoroughly so no fertilizer burn occurs. Apply 10-10-10 no more frequently than once a month and no closer to the first-frost date in fall than 6 weeks.

The same procedures apply when using organic fertilizers. Cottonseed meal, finished compost, or aged manures can all be used effectively. If compost is still "working" (i.e., warm to the touch 6 inches inside the pile) or manure is fresh, it should be applied at least a few weeks before planting, mixed with the soil, and allowed to finish any heating before planting takes place.

About 6 weeks before the first-frost date, increase the time between seedbed irrigations; if, for example, you were irrigating every 7 days, increase the time interval to 10 or 14 days. However, don't let the plants suffer prolonged wilting from droughtlike conditions. Midday wilting is normal on a hot summer day. Irrigation should be reduced, not stopped, during dry fall weather.

Plants will naturally prepare themselves for winter as autumn days get shorter and nights become cooler if you grow them drier. Native plants should be able to withstand winter where they are normally hardy with only a little help from gardeners. Too often, in their eagerness to grow a plant as fast as possible, gardeners overfertilize and overwater late in the season, preventing the natural conditioning a plant goes through to get ready for winter.

Transplanting. Container-grown seedlings can remain in pots only a limited time before they grow too big. I like to begin transplanting small-seeded plants to larger pots when they have four true leaves. (Figure 5 shows the difference in appearance between a true leaf and the seed leaf, or cotyledon.) Seedlings grown from large seeds can usually be transplanted as soon as they have only two fully expanded true leaves. The size of the pot you transplant into will depend upon both how fast your native plant is expected to grow and how long you plan to keep it in the pot. The number of seedlings you plant per pot is purely a matter of personal choice. If you

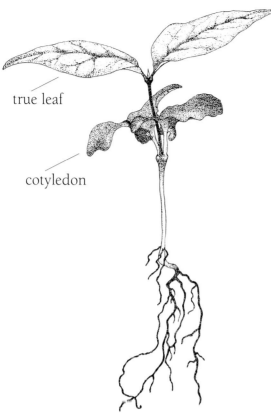

Figure 5. Seedling, with cotyledon and two true leaves.

want a clump of trees or shrubs, the time to start the roots growing together is at the first transplanting. If you change your mind, you can always remove extra plants after you know how many survive.

I like to grow transplants in a potting mix that is similar to the seeding mix I used. The 3:1 pine bark–peat mix mentioned earlier works nicely. If you are growing trees or taller shrubs in containers, you may want to add up to one part coarse sand to your potting media. The weight of the sand will help to anchor plants and prevent their blowing over.

When transplanting, work in a shady location. Knock your seedlings out of the pot by turning it over and rapping the rim on a hard surface. For pots in which seedling roots have barely reached the bottom, gently patting or squeezing the bottom of the pot may be all it takes to dislodge the seedlings (see Figure 6). However, I usually find it necessary to rap the pot rim sharply on some hard surface in order to loosen the seedlings. *Gently* pull seedlings apart, holding them by either the seed leaf or true leaf but *not* the stem. Replant seedlings into moist potting media at the same depth at which they were growing in the seeding container and then *gently* irrigate to firm media and eliminate any air pockets around the roots. Do not press growing media around the roots. Pressing the media too firmly may break tender young roots or create media that is so compact young roots will penetrate it slowly if at all. Be careful not to let the roots dry out at any stage of the transplanting process.

Place recently transplanted seedlings in the shade to recover. Move them back into a sunny location at dusk or on an overcast day. Do not fertilize for at least 1 week after transplanting. Avoid pruning or pinching seedlings for 2 weeks before and 2 weeks after transplanting because this may retard new root growth.

Seedlings grown outdoors in the ground should be transplanted only when dormant. In the coastal plain and much of the piedmont, this is in mid- to late winter. In the mountains, transplant in late winter or very early spring, as soon as possible after the soil thaws. Most seedlings make better transplants if they are grown in outdoor beds for two years. This creates larger, tougher plants that are able to grow without pampering when they finally enter the garden.

Winter Protection. The roots of plants are much more sensitive to cold-weather injury than the tops. Normally, roots are in the earth, which is much warmer than the air during cold weather. If you are growing plants in containers, the roots of the plants can be injured when the containers are exposed to cold air because growing media is exposed to the air on all sides. Therefore, special care must be taken to keep these plants alive by protecting the roots in hardiness zones colder than 9 and the warmer parts of 8.

The easiest way to protect container-grown plants from winter cold is to transplant them into the garden in the fall. After the first few fall frosts, mulch in the mountains and piedmont. Once roots are in the earth, geothermal heat will keep them warm naturally. Containers holding plants not transplanted into the earth should be protected from severe cold as well as from large changes in temperature. One of the easiest ways to accomplish this is to move plants in containers to a shady location in November and totally cover the pots, including the media surface, with organic mulch (see Figure 7). The mulch will prevent cold air from penetrating spaces between pots. Move these plants back to a sunny growing area in spring after the danger of temperatures below 25° has passed. Your native plants should be able to withstand frost with little damage. However, you may notice a dramatic yellowing of leaves shortly after container-grown plants are moved back into the sun. This is a condition nursery growers call "frost burn." No permanent damage has occurred; plants will turn green again once warm weather returns and you resume your

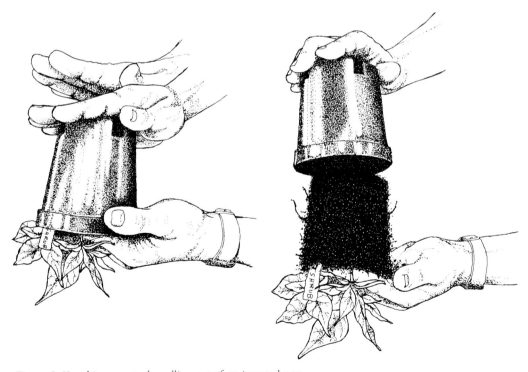

Figure 6. Knocking a potted seedling out of an inverted pot.

regular fertilization and irrigation schedule. If you will be growing plants in pots a second season, begin fertilizing at a light rate a few weeks before the average last-frost date in the spring.

The first year bed-grown plants or fall-transplanted seedlings have to endure winter conditions, they should be mulched. I prefer pine needles or clean oat straw to other mulches. Hay usually contains weed seeds, pine bark may float away in winter rains, and hardwood leaves compact too much. Completely cover the soil surface with mulch only after the soil has cooled in the fall, generally in mid- to late November. The purpose of the mulch is to prevent fluctuation in soil temperature and the resulting heaving, not to keep the soil warm. Mulch should be removed in spring after there is no longer a danger of temperatures below 25°. Plants to be grown a second year should be fertilized at the maintenance rate (see Appendix 1) before buds begin to swell in the spring. In most areas, this is 4–6

Figure 7. Container-grown plants being covered with mulch for winter protection.

weeks before the average last-frost date. Plants that are to be removed from seedbeds in late winter should not be fertilized before transplanting.

Cuttings

Asexual propagation involves many, many techniques, including budding, division, grafting, layering, marcottage, rooting cuttings, and tissue culture. Even propagation by cuttings can be divided into categories based on the part of the plant from which the cutting was taken (such as leaf, root, or stem) or on stage of growth, time of year, and treatment techniques. Asexual propagation, while one of the most rewarding of gardening activities, can be complicated. I'll try to keep it simple. In this book, the only methods I'll describe are propagation by stem and root cuttings. If you wish to try other forms of asexual propagation, consult the relevant books listed in the Selected References.

Why would anyone want to root a cutting? The most common reason is to get an exact copy of the parent plant. When you grow plants from seeds collected in the wild, the plant you grow will not be exactly like the parent. Your seedling may be similar to the parent, but it will not be exactly the same. Nor will any seedling be exactly like any other. Seeds I collected from one flame azalea (*Rhododendron calendulaceum*) grew a population of plants that had flower colors ranging from deep red-orange, through melon tones, to golden yellow. If I had asexually propagated the parent bush, the plant grown from the rooted cutting would have had flowers colored bright orange.

Another reason to propagate via cuttings is that sometimes asexual propagation is the only means available. For a variety of reasons, a desirable native plant may never set seeds or you may not be able to collect the seeds. For example, suppose the plant you covet for your home garden is at your summer place; you can root stem cuttings in the summer even if you can't be there in the fall to collect seeds. Or perhaps you covet a particularly striking *Chionanthus* with large, showy flowers. The showiest *Chionanthus* flowers are on male plants, and since male plants don't produce seeds, to duplicate this plant you must propagate asexually.

Professional plantspeople propagate via cuttings so they are sure that the plant they are selling is exactly the same as the parent. Wouldn't you be disappointed if you bought a dwarf plant and it turned out to be tall, particularly if you had planted it under a window? Furthermore, it is often possible to have a flowering plant much sooner from cuttings (see Table 1). I've rooted softwood Carolina silverbell (*Halesia carolina*) cuttings one summer and had flowers on the young plants 21 months later. Seeds of Carolina silverbell have a double dormancy, so they usually take two years just to germinate. Woody plants often remain juvenile for a few years after seed germination, just as humans can't reproduce for some years after they are born. Juvenile woody plants tend to hold onto their leaves into the winter and do not flower. For this reason, Carolina silverbell seedlings usually do not begin to flower until they are 3 years old, 5 years after their seeds were sown. That's too long for most nurseries to wait, so they root stem cuttings one summer, grow plants into small trees the next summer, and sell the trees in

Table 1. Comparison of Time Required to Grow Carolina Silverbell Trees from Cuttings and from Seed

Means of Propagation	Year 1		Year 2		Year 3	
	Spring	Fall	Spring	Fall	Spring	Fall
Cuttings	stick cuttings	cuttings root	rooted cuttings grow		small trees flower	
Seeds		collect seeds	plant seeds	warm, cool stratification	seeds germinate	flowers in 3–5 more years

flower the following spring, less than two years from the time they took the cutting (assuming that everything goes right).

The greatest potential drawback to growing plants from cuttings is also the biggest asset. You know exactly what you are getting. Some of the fun involved in sowing seeds is seeing the differences you get. You can count on some seedlings not being as attractive as the parent, but you also have the chance to grow the best specimen there has ever been or something highly unusual. On the other hand, if one branch on a green-leaved plant has purple or yellow leaves and you plant seeds that came from that branch, all of the seedlings will probably have green leaves. If you root a stem cutting from the branch with purple leaves, the whole plant grown from the cutting is likely to have purple leaves. If you propagate this plant from root cuttings rather than stem cuttings, however, it will revert to the dominant characteristic of the parent—that is, green leaves. Once you know what you want to accomplish, choosing your propagation technique is easier. There is no reason why you cannot use both techniques for the same species. Ornamental plant breeders will often control pollination of plants, grow a population of seedlings resulting from their controlled crossing, and then asexually propagate the best plants.

Stem Cuttings

Getting Ready. When propagating plants from stem cuttings, you are removing a part of the desired plant and providing this rootless plant part with conditions that will keep it alive until roots develop.

Plants obtain nearly all their water and essential nutrients through the roots. When you separate a plant from its roots, it will die unless special precautions are taken. Therefore, it is important to have everything ready *before* you collect your cuttings.

You need to provide some means to keep

the cutting from drying out. Leaves will continue to lose moisture after you separate the cutting from the mother plant, but the ability of the cutting to replenish this moisture to the leaves is limited. Growers in nurseries use fog-generating machines, electronically controlled fine mist, and other sophisticated techniques to keep the humidity high around plant leaves while waiting for roots to develop. On a small scale, the same thing can be accomplished for a lot less cost by covering the above-ground portions of a stem cutting with a polyethylene plastic bag (such as a bread wrapper or food storage bag), provided it is protected from direct sun so that the bag does not become a solar oven and cook rather than coddle the cutting.

Green plants will have some stored food in the stem cutting, but most cuttings taken from new plant growth have very limited food reserves. Usually, the earlier in the growing season you take a stem cutting the softer it will be. These "softwood" cuttings contain little in the way of food reserves. Plants must use stored food to manufacture roots. If there isn't enough stored food, they must manufacture more or no rooting will occur.

Green plants make food through the process of photosynthesis. Photosynthesis requires light, but light energy can quickly become heat energy, by the same process that makes us warmer when we sit in the sun. When plants get warmer, they lose moisture more rapidly. Your job as a plant propagator is to balance temperature, moisture, and light so your cutting stays alive long enough to develop roots and be able to fend for itself.

Most woody plant cuttings require conditions that are a bit more exacting than the "slips" gardeners have been sharing for years. Putting stem cuttings of most shrubs and trees in a glass of water in the east window won't produce roots unless you are working with a very few easily rooted plants like pussy willows (*Salix sp.*). Even pussy willows will root in higher percentages if you follow the general suggestions in this book.

Propagating Media. So far, I have been discussing the part of the cutting that is already grown, that is, the stems and leaves. What about the part that grows the roots? This part is stem also, but it will be stuck into a dark, moist place where it will be surrounded by soillike particles and air. The proportion of air to soillike particles will often make the difference between success and failure when you try to root stem cuttings. This mixture is called the propagating medium, or just media, by horticulturists. It is one of the most popular topics for discussion at nursery meetings.

I've found that most woody plants will root in a mixture of sphagnum peat and perlite. I like to mix them in equal parts; in other words, for every bucket of peat, I mix in one bucket of perlite. I moisten the peat, remove it from the bale, and crumble it with my hands. (The easiest way to moisten peat is to use the following procedure. Cut a hole in the plastic covering near the top of the bale and insert a hose in the hole. Turn on the hose and let water run *slowly* into the bale until it comes out the hole; then turn the water off. After a couple of hours, poke holes in the bottom of the plastic wrapper to let excess water drain away.) Do not rub peat through a screen. This will create very fine, dustlike peat that is hard to get uniformly moist and doesn't have enough air space to allow newly formed roots to breathe. Perlite should also be moistened before mixing, as dust will fly everywhere when you pour dry perlite. Some folks like to propagate in medium to coarse vermiculite or coarse sand. For many plants these media are excellent. However, avoid fine grades of sand or vermiculite because they won't have enough air spaces for young roots. When there isn't enough air in the rooting media, roots and stems will eventually rot. If you choose to propagate in pine bark, sift the pine bark through a $1/4$-inch-mesh screen to get it fine enough. Most pine bark is too coarse to be used as a rooting medium when you get it and, as a result, won't hold enough moisture around the base of your cutting, allowing it to dry out unless the whole pot, top and bottom, is enclosed in a plastic bag. None of these media components contains soil, but all are originally naturally occurring materials. Perlite, sand, and vermiculite are minerals. Peat and bark are plant products. Nevertheless, when mixed together they are often referred to as "artificial" media.

Sanitation is very important when rooting cuttings. A cutting is a plant part struggling to survive. In rooting cuttings, we place a recently cut stem into a warm, moist area perfect for disease growth, but expect the plant to survive without becoming diseased. You can help prevent disease by keeping things as clean as possible. Nurseries generally do not bother to pasteurize peat, perlite, pine bark, or vermiculite. If sand is mined from the earth rather than dredged from a river, it is usually not pasteurized either. However, if you are using sand that came from a river or stream or that you dug out of the yard, put it in a pan and pour boiling water over it before mixing it with other media components.

It is also important that containers be clean. New clay or plastic pots do not need to be pasteurized. However, most of us reuse plant-growing containers. Dirty or used containers should be washed thoroughly. Be sure to remove all clinging soil; then dip the containers in a mixture of one part chlorine bleach to nine parts water and allow them to air dry. The bleach solution should kill disease organisms that survived the pot washing.

Any container in which plants will grow is suitable for rooting cuttings. Once a stem cutting develops roots, you will need to keep it in that container for a few weeks until the cutting has grown enough roots to support the plant without the special moisture and light conditions provided while rooting. Therefore, your container should allow water to drain away from roots, be substantial enough not to come apart, and contain no potentially harmful chemicals. I prefer clay or plastic for these reasons. Treated wood or cardboard hasn't worked as well for me. However, naturally rot-resistant woods, like cedar and cypress, have been excellent and lasted for years when I've been lucky enough to find containers made of them.

Rooting Hormones. Most professional growers apply synthetic rooting hormones to the wounded surface of stem cuttings as soon as possible after making the wound. Hormones can increase the speed of rooting and the number of roots formed. However, hormones are not essential to rooting most cuttings. The most easily obtained rooting hormones contain IBA (indolebutyric acid) mixed with talc. The concentration of hormone you want to use will depend upon the type of cutting you are taking as well as the plant you are trying to root. I've had greatest success on semihardwood cuttings with powders containing .8 percent (8,000 parts per million [ppm]) IBA, such as Hormodin No. 3. If you choose to apply a liquid formulation of IBA, apply it as a "quick dip." This means the cutting is dipped into the liquid and immediately removed. The whole process has an elapsed time of about one second. I've gotten results with a .25 percent (2,500 ppm) IBA solution similar to those achieved with .8 percent powder because the liquid enters the cutting more efficiently. This greater efficiency means that a lower concentration of hormone is needed when using liquid formulations than when using powders. See Appendix 7 for the names of some rooting hormone suppliers.

Another synthetic rooting hormone is naphthaleneacetic acid (NAA), which is found in many formulations of the commercial preparation Rootone. Generally, I've found IBA more effective when rooting woody plants and NAA better for herbaceous plants. The preparation available locally that I usually suggest for new propagators is a powder (Chacon Rootone) with both IBA and NAA in concentrations suitable for softwood and early semihardwood cuttings. It is conveniently packaged in a zip-top plastic/foil pouch.

If you are a label reader, you will notice that fungicides, boron, and a number of other things can be added to the propagator's magic rooting elixir. IBA is customarily all I use with woody plants. If you do everything else properly, it's probably all you will need to root stem cuttings for most plants in this book.

When using rooting hormone, remove a small amount from the can or pouch of powder or the bottle of liquid. Tightly reseal the container and store the unused portion at room temperature or in the refrigerator, not where it will get hot. Handled this way, your supplies can't become contaminated from contact with plant parts and your hormone liquids and powders should keep their effectiveness for years.

Depending on the size of the cutting, I usually treat about $1/4$ inch of the stem base with hormone. Larger cuttings have a slightly greater area treated. This area and the stem just above the treated area are where most of the roots will appear. Make a fresh cut on the stem immediately before treating the stem with hormone. If the cut surface is allowed to dry before hormone treatment, the effectiveness of the hormone will be greatly reduced.

Types. Only certain parts of plants will root at certain times. Knowing when to take cuttings is just as important as knowing how to take them. Three basic types of stem cuttings are mentioned in this book. (1) A softwood or greenwood cutting is taken from new wood that is just beginning to harden or become firmer. Softwood cuttings are very succulent and bend or snap easily, with no bark clinging to the stem. This is the most common type of spring cutting. (2) Semihardwood cuttings are the type most commonly used in the nursery industry. These are taken from wood that is partially matured but still easily snapped. With this type of cutting, bark will often cling to the stem after the stem has been snapped. These cuttings are usually taken from late spring into early autumn. (3) Hardwood cuttings are taken while the plant is dormant from late fall through early spring. Hardwood cuttings from deciduous plants should be taken after leaves have fallen.

All cuttings should be taken according to stages of plant growth rather than according to the calendar. There is too much variation from one season to the next to rely on the calendar. Guidelines given in the plant list are a place to start experimenting. Experienced propagators can discriminate very small differences in the look or feel of plant tissue—too soft or too hard—used for cuttings. This can be the difference between success and failure in some difficult-to-root species. Experience takes time. Until you gain experience, find comfort in knowing that I or my friends succeeded, at least once, using the suggestions in this book.

Gathering. Cuttings should be collected from healthy, normally growing plants. If possible, avoid cutting wood that appears to be diseased, infested with insects, or undergoing distorted growth. Stem cuttings should be taken from the parent plant when the plant is full of water (turgid), not wilted. For this reason, the best time to take stem cuttings is early in the morning while temperatures are still cool and dew remains on the leaves. Check the parent plant the day before you intend to collect your cuttings. If it is wilted, irrigate and then take cuttings a day or two later.

As soon as cutting wood is removed from the parent plant, place it in a dark, not clear, plastic bag. Once you have sufficient cutting wood, close the bag and remove it from direct sunlight. If more than a few minutes will elapse before you prepare and stick cuttings, place them, still in the bag, in a portable picnic cooler with ice. The plastic bag will keep the humidity high around the cutting. Efforts to keep the cutting cool will be rewarded with a higher percentage of rooting later. Do not, however, freeze cutting wood. Any further preparation of cuttings should be done out of direct sunlight and wind to keep moisture in the cutting. If cuttings must be stored, store them somewhere cool and moist. Keep storage time to a minimum. For most shrubs and trees, the sooner a cutting can start rooting, the higher your chances for success will be.

Much has been written about where on the plant you should take the cutting. On wild plants, or those in the landscape, often you have no choice. The only healthy, vigorous growth may be on one side of the plant. Given a choice, I prefer to collect my cutting wood from the top to middle of the plant or shrub, where it has been receiving as much sunlight as possible. This is because I want the stockiest growth for the cutting. For most plants the differences in percent rooting between cuttings taken from the shade or sun, east or west, side of the plant are minimal. If you're an average gardener you want only a few plants, so stick twenty cuttings and then select the best ten to pot after they are rooted. Throw away those that are slow to root or not as vigorous.

Preparation. Sanitation is important at all stages of plant propagation. However, it is especially important when preparing cuttings because you are wounding the plant, and the open wound provides a perfect entry for disease if your working area, cuttings, and tools are not clean.

The tools you'll need to prepare cuttings are the same ones used in collecting the cutting wood: a sharp knife and a pair of hand pruners (secateurs). Sharp cutting tools are most essential at this stage since you want to leave a clean wound rather than one that is ragged or surrounded by crushed tissue. If you suspect that disease may be transmitted by a dirty tool, disinfect the tool by dipping the blade in rubbing alcohol or the 9:1 water-bleach solution mentioned earlier for cleaning pots. Disinfect tools only when absolutely necessary since disinfecting solutions can be corrosive to metal tools and harmful to plant tissue.

If cutting wood appears to be dusty or dirty, wash the cuttings in a mild solution made from a liquid dishwashing soap mixed with water. Thoroughly rinse leaves and stems with clean water immediately after washing the cuttings in the soap solution. This sanitation procedure is usually more important for cuttings taken from the wild than for those taken from landscape or nursery stock plants. Wild plants often have insects and insect eggs on their leaves. If these aren't washed off, an unexpected critter population may appear after a few weeks in the propagation area.

Branches or any wood that has been collected could be used for a stem cutting. However, young or juvenile tissue is most likely to root. Even on dormant hardwood cuttings, wood that grew the previous year is more likely to root first and have a more robust root system. Fortunately, the young wood that will become the cutting is at the end of the branch or twig.

The length of your cutting may be predetermined for you. I prefer 4- to 6-inch-long cuttings. But if the parent plant only grew 3 inches, your cutting has to be smaller. Working quickly, remove all but the two to five leaves at the tip of your cutting. Make a new cut across the basal end of the stem. If wounding has been suggested for the particular plant you are propagating, remove a shallow slice from the base of the stem, barely cutting through the bark (see Figure 8).

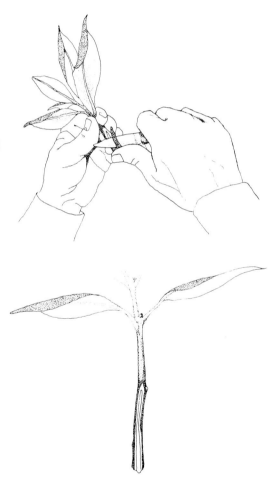

Figure 8. Wounding a stem cutting.

Wounding is necessary for some species in order to develop a larger root system on hardwood and semihardwood cuttings. In these species, roots will often develop only on the wounded area at the base of the stem. To increase the number of roots, the size of the area wounded is increased.

Except for cutting the base of larger stems, this whole procedure is usually done with a knife. I prefer cuttings from the tip of a branch or twig whenever possible. Cuttings taken from more mature wood further back on the stem will usually take longer to root as well as root in lower percentages.

Sticking. Procedures for sticking cuttings will differ slightly depending upon whether or not you used hormone and, if you did, which formulation you chose. If you did not use hormone, simply insert the cutting, stem end down, about an inch into the rooting media. Be sure the base of the cutting is at least an inch above the bottom of the container. The media in the bottom inch of any container will be wetter than that above. Avoid it.

Cuttings should be placed far enough apart so that leaves from one cutting do not overlap the neighboring cutting. On large-leaved plants like rhododendrons, you may need to cut the leaves in half to provide enough room between cuttings (see Figure 9). Cuttings are generally stuck 2 to 4 inches apart if they are being stuck in a community flat. If you are sticking cuttings directly into small pots where they will be allowed to grow after rooting, easy-to-root plants are usually stuck one cutting per pot. Hard-to-root plants or those that usually grow in clumps or thickets are often stuck with two or three cuttings per pot.

If you used a liquid rooting hormone, let the hormone solution dry before you stick

Figure 9. Stuck rhododendron cuttings, with leaves cut in half to avoid overlap between cuttings.

the cutting. Since most of these solutions consist of a hormone dissolved in alcohol and then diluted with water, drying time is relatively short, no more than a minute or two. If you used a powder formulation of hormone, open a hole in the media with a dibble such as a pencil or knife and then stick the cutting. This prevents rubbing the hormone powder off the cutting when you insert it into the media. Gently firm the media around the base of the cutting and water so that the media comes in intimate contact with the base of the cutting and no air pockets exist.

Care during Rooting. An experienced propagator can prepare and stick hundreds of stem cuttings in less than five minutes. For the amateur, speed is not important. Doing things properly is. Make sure you follow each step carefully. You may not get a chance to work with some plants for another year, or ever. Things need to be right the first time.

For this reason, I suggest that you stick only a few minutes' worth of cuttings at a time and then water them in. Delay in watering may allow cuttings to lose enough moisture to significantly lower rooting percentages or even to prevent rooting. As soon as possible after watering-in, cuttings should be put in their poly tent or under mist so they do not dry out.

Mist systems should be set so that media stays moist, but not sopping wet. Generally, they are on a time clock that will turn the misting nozzles on an hour after sunrise and turn them off an hour before sunset. This allows leaves to be drier during the night, when humidity is naturally higher. Dry leaves are less likely to develop disease.

The interval between times when the mist is being applied can be regulated by a variety of means. If the regulating system depends on a time clock, set the clock for the minimum interval that will do the job. Too often, mist systems keep media overly wet, and cuttings rot as a result. Commonly, mist will be on once every 4 to 6 minutes shortly after plants are stuck; then intervals will be increased gradually until mist is coming on only every 10 or 12 minutes on sunny days. On rainy or cloudy days, it often is not necessary to run mist at all.

One of the simplest, most inexpensive, small-scale propagation systems I have seen employs small, six-pack styrofoam coolers. Drainage holes are punched in the bottom of a cooler with a pencil or screwdriver, media is poured in to a depth of a few inches and moistened, and then cuttings are stuck. If you prefer, cuttings can be stuck in individual media-filled and moistened pots and then placed in the cooler (see Figure 10). A polyethylene sheet is stretched over the top; then the whole propagation chamber is placed in a partially shaded location until cuttings are rooted. Once they are rooted, cuttings are "hardened off" (exposed to lower-humidity air) by poking a few holes in the plastic top one day, making a few more holes a couple of days later, and finally removing the plastic top in a week to ten days. These cooler rooters are easy to clean with soap and water and disinfect with a bleach solution before propagating the next crop of cuttings.

Most softwood and semihardwood cuttings will root best when the temperature of the media is 70–75°. Artificial warming of propagating media is called "bottom heat."

Whether you choose to use bottom heat or not, no supplemental heat is needed during the summer unless nighttime temperatures are regularly below 60° or temperatures during the day rarely reach 80°. A warm temperature around the roots will also help hardwood cuttings to root faster, but it is important that the temperature around the top of hardwood cuttings should be cool, not warm. For small-scale propagators, providing conditions that will keep the bottoms of hardwood cuttings warm while enabling the tops to remain cool is too costly. Since warm temperatures around the stem base help roots form faster but are not essential to root formation, most amateur growers simply wait longer for their cuttings to root and avoid the bother of bottom heat.

Moisture control is just as critical as temperature control when rooting cuttings. If you are using a plastic tent, once you close the tent, try to leave it closed. Every time you open the tent, humidity escapes. You do not need to water the media since no moisture should be escaping from your closed system. Open the tent only if you are un-

Figure 10. Cooler rooter.

able to control temperatures adequately. When temperatures inside the tent rise above 100°, ventilate to release trapped heat. If media starts to dry out, irrigate, let the media drain, and close the tent again. You are trying to keep high humidity around the leaves to prevent water loss from the cutting while keeping the right balance of air and moisture in the rooting media to encourage root development. Adding water to media when it is not needed will keep the area around the base of the stem too wet, which can cause stems to rot.

Timing. I've mentioned all the things you can do to help cuttings root faster: provide warm temperatures, maintain a proper moisture balance, treat with hormones, and have clean, healthy cuttings to start. If one of these factors isn't quite right, roots will take longer to develop if they develop at all. Rooting cuttings is a race to keep the plant alive until it can support itself.

If you have done everything right, you can expect root development to begin in a few weeks on easy-to-root plants. I've rooted semihardwood cuttings of deciduous hollies in 3 weeks and had them putting on new growth in 6. On the other hand, I've worked with cuttings from a naturally dwarf *Rhododendron maximum* that took 5 months to show the first roots under ideal conditions. Different plants have different requirements for growth, different growth rates, and different rooting rates. Since the time required for rooting varies from plant to plant and season to season, even under similar conditions, I haven't indicated rooting times in the plant list. For most plants, I suggest that you don't touch the cuttings for at least three weeks after sticking.

When you feel some resistance as you tug gently on the top of a cutting, you can be fairly certain that roots are starting to anchor the cutting into the media. A week or so after you first feel this resistance, start reducing humidity by removing the plastic covering early in the day and replacing it by midmorning or by increasing the interval between mistings. You can also poke holes in the plastic, as described for the cooler rooter. Within 7 to 10 days you should be able to remove the plastic or turn off the mist entirely. I prefer to remove the plastic or turn off the mist in late afternoon so that the plant goes through the cooler temperatures of evening and the naturally reducing humidity of the next morning to begin hardening off.

Care after Rooting. Once cuttings have rooted, most folks want to fertilize them or transplant them into the garden. For a very few woody species, this would be okay. For most, these plants are still babies and still need special care.

For many flowering shrubs and trees, including deciduous azaleas, dogwoods, and silverbells, if the cuttings do not begin new shoot growth before you disturb the roots, they will probably never develop into the plants you want in your landscape. Often, within a few months they will be dead. Many of these plants require long days (or short nights) to put on new growth. Nursery growers who take cuttings in early spring or late winter from container-grown plants that have been forced into growth in heated greenhouses have their cuttings rooted when days are naturally long so they begin new growth on their own. The way other commercial growers get this type of rooted cuttings to grow is to expose them to "long days" by stringing lights over the cuttings. Using these lights, day length is extended to about 18 hours to force new growth. Once shoot growth starts, the plants can be handled normally. One disadvantage of this technique is that plants are often in the middle of a tender flush of growth just as the first frosts are about to occur. Tender new growth is susceptible to damage by cold, so growers must protect these cuttings from low temperatures during the first winter. I think this is too much trouble for most gardeners.

A technique that works just as well for this type of plant but requires far less precise methods and no more winter protection than you provide for any potted plant is to move the hardened-off, rooted cuttings, still in their *undisturbed* rooting medium, to a cold frame or shady area where they sit under normal care for the rest of the summer. In the fall they go dormant and receive the same sort of winter protection that was suggested above for seedlings. The next spring, as soon as new leaf growth starts, the cuttings can be transplanted to containers or nursery transplant beds, fertilized, and watered like other plants. I've found that this technique works well for me, particularly when I need only a few rooted cuttings of lots of different plants. I can stick the cuttings in small pots, being sure to label each pot, and then treat all the cuttings the same. The trays (flats) I use most often are 18 inches square. These trays will hold 36 square 3-inch pots in 6 rows of 6 pots each. This makes it easy to handle 6 pots of rooted cuttings for 6 different species at the same time.

Root Cuttings

Root cuttings seemed to exist in the realm of magic when I was first introduced to them. Perhaps this is because in this type of propagation cuttings are taken from beneath the earth when the woody plant is dormant and then placed back into dark regions to perform their miracle of regeneration. Growing root cuttings is just as much an act of faith as planting a seed or sticking a stem cutting, but the aura of mystery surrounding root cuttings exceeds that surrounding all but grafting in the world of macropropagation.

Fortunately, keeping just a few things in mind will usually ensure success with root cuttings. No special structures, bottom heat, plumbing, plastics, or hormones are needed. However, you do need a bit more than faith or blind luck to succeed.

Getting Ready. Timing, a sharp knife or pruners, and moisture management are the keys to success with root cuttings. All the plants in the plant list for which I have suggested root cuttings should have the cuttings taken while the plants are dormant. Unless otherwise indicated, cuttings should be taken in December or early January so that there is plenty of time for buds to form and new roots to begin growing from the chunk of root before the warmth of spring.

When you take a root cutting, the buds that will become your new plant may already be present, or they may need to form. Forming new buds takes both time and energy. Buds are most likely to form on the current season's roots. The energy for bud formation is stored in the root pieces that become your cuttings. For this reason, it is important to get stocky root pieces, close to pencil thick if possible.

In the few commercial nurseries that still employ root cuttings for propagation, plants are root pruned by cutting with a sharp spade forced straight down into the soil 6 to 12 inches from the parent plant in late winter; then the soil just beyond the pruned roots is mechanically cultivated. During the growing season, new, vigorous growth will develop from the cut portion of the old roots, extending into the cultivated soil. These new, vigorous roots become the root cuttings. They are easily located because you know where you cut to root prune and the cultivated soil is usually still soft. A bonus from this technique is that roots are usually straight and a uniform size. Another technique for obtaining root cuttings is to work with container-grown stock. The potting mix is washed off the roots, cuttings are taken, and then the plants are repotted. These cuttings are rarely straight or uniform but grow perfectly good plants.

Media. Artificial media used for stem cuttings can be used for root cuttings as well and should work best for you if you will be propagating root cuttings in containers. I prefer to add 10–20 percent coarse sand to any media used for propagating root cuttings and have used pure sand successfully for some species. However, root cuttings can also be propagated in cold frames or at the edge of the garden in well-drained soil. The media should be uniformly moist when you stick your root cuttings.

Gathering and Preparation. Dig wherever it is necessary to obtain cuttings of last season's vigorous growth. For an established tree, this is usually a foot or more beyond the drip line and within a foot of the soil surface. Fortunately, most of the plants propagated by root cuttings are large shrubs or small trees with vigorous root systems, so finding these roots is usually not difficult, even if it is often a cold and muddy task.

Your objective is to obtain propagating material, not to injure the tree, so collect only as much as you need. Then fill the hole and firm the soil. Your rooting percentages are likely to be high with this technique, and your cuttings will generally be only 2 to 4 inches long, so you don't need to take much cutting material. Four or five large roots, approximately pencil thick, are usually plenty.

Fibrous roots and soil should be removed from root pieces in the field; then the cuttings are put in a clean polyethylene bag, placed out of the sun (in your jacket pocket?), and transported to a location away from the drying effects of direct sun and wind where the cuttings can be prepared.

Maintaining polarity—that is, keeping track of which end is up and which is down—is very important with root cuttings. The end closest to the parent plant is called proximal or "up." This should always be the top of your cutting. Since uniform root pieces often look alike, the way most nursery growers keep track of up and down on root cuttings is to cut the root straight across (i.e., perpendicular to the way it was growing) on the proximal end and at a sharp angle like that of a splitting wedge at the bottom or distal end. When proceeding in this way, you should make the first perpendicular cut when you collect the cutting and then follow through with perpendicular and angled cuts as you divide the collected roots into smaller pieces. Unless otherwise advised, make your cuttings about 3 inches long.

I was taught that root cuttings must not be allowed to dry out, but that all cut surfaces should be allowed to dry. Perhaps the best way to do this is simply to prepare your cuttings and leave them exposed on a clean bench or in an empty casserole dish for a few hours, as you might do when preparing potato seed pieces. Some nurseries dust cut surfaces with fungicides, but fungicide treatment is not absolutely necessary if you need only a few plants, allow cut surfaces to dry, and manage moisture well.

Sticking. There is nothing complicated about sticking root cuttings. Place them in rooting media in such a way that their tops are just below the surface and their bottoms are at least a couple of inches from the bottom of a well-drained rooting medium; then gently firm the media around the cutting.

I believe in sticking root cuttings vertically, with the flat end on top, although I have seen it written and heard it argued that if you lay them on their sides and then fill the container with rooting media, it won't matter whether you know up from down on your cuttings. The difference between sticking cuttings vertically or burying them horizontally can be a 50 percent or better improvement in your success in rooting. Vertically stuck cuttings always have been more successful for me, with anywhere from 85 to 100 percent of them rooting for most species. I once did a comparison of horizontal and vertical root cuttings. After the horizontally stuck cuttings had just lain there dormant for three months and most of the vertical cuttings had sprouted new growth, I turned half of the horizontal cuttings to vertical. Those were the only ones to break bud and root.

Care. Once cuttings have been stuck, according to what I was taught, the tops of the containers should be dressed with half an inch of clean sand. This is to prevent the development of mold or moss, and it works. Do not water-in root cuttings as you would stem cuttings. You should have been using moist media when you began. The biggest danger to a root cutting is rotting from excessive wetness.

Containers of root cuttings are usually put in a cool, dark location. I've kept them in the basement, the garage, and a hole under the deck used for forcing pots of bulbs, for example. If the location is outdoors, I normally put an old metal screen over the cuttings to discourage inquisitive squirrels and chipmunks, and I check occasionally to see if they are drying out. Avoid locations where the root cuttings will freeze solid. If you put these containers in the right location, December root cuttings won't need watering until spring.

I have not had good luck with root cuttings in warm places or the greenhouse. Buds on these cuttings form and break ahead of new root growth. If the temperature is warm, these new leaves and shoots can lose moisture faster than the root cutting can supply it, so the new plant dries out and dies. If you plan to try root cuttings in the greenhouse, choose a cool greenhouse and shade the rooting container, or put it under a bench away from heat pipes.

Fertilization. Your rooting media contains virtually no fertilizer. Plants need nutrients to grow. Therefore, when do you begin to fertilize your cuttings?

Stem cuttings that root rapidly and will start new growth the same season they are rooted should be fertilized as soon as they can use the fertilizer, which is as soon as they have roots. Remember, though, that new roots are very sensitive to fertilizer, so you should treat them gently. Fertilizing with water-soluble fertilizers like Miracle-Gro or Peters at one-half the rate suggested on the label, as described for seedlings, works well. If plants will remain in the containers where they were rooted over the winter, apply fertilizer solutions up until 6 weeks before the expected first-frost date in fall. Low rates of slow-release, dry granular fertilizers like Escote and Osmocote may also be used. Be sure to read the label: these slow-release fertilizers are designed to feed the plant over a certain length of time. If your first frost is expected 4 months after you fertilize, use a 3-month controlled-release fertilizer, not an 8-to-9-month fertilizer. On newly transplanted liners, never use more than the lowest suggested rate on the fertilizer label. Too much fertilizer can kill a plant. Too little will just cause the plant to grow more slowly, not kill it.

Root cuttings should be fertilized only after new root growth has begun. New stem growth may be apparent in late winter or early spring, but new root growth may not begin for another 4–6 weeks, so please be patient. Mid- to late spring is the time to think about lightly fertilizing root cuttings. The cutting pieces should have enough stored food to sustain the young plant until then.

If you wish to use natural organic fertilizers, be sure you use finished rather than still-active compost or aged manures. At this delicate stage of growth, I prefer to top dress with cottonseed meal or weekly applications of "deodorized" fish emulsion diluted in water. Plants that grow a little more slowly are far better than plants that are damaged or dead because I was too eager and used too much fertilizer.

Plants that must go through the winter chilling process or artificial lighting before they begin to grow should not be fertilized, or should be fertilized very lightly once rooted. Research has demonstrated that fertilizing these plants before they start to grow can increase the chances of their dying during the winter.

Pruning and Transplanting

You will want most trees to have a single trunk. Therefore, no pruning of the central leader or top of seedlings or rooted cuttings is desirable unless you have more than one top. If your plant has multiple leaders, remove all but one. Choose the one with the thickest diameter to be your leader. That's the way it would have worked in nature, even though the leader selected may not have been tallest when you pruned it. Sometimes many sprouts will arise from the base of a single stem. If you are growing a single-stemmed tree, remove all of these suckers as soon as they are noticed. If you are growing a tree that might normally appear in clumps, leave as many stems as you wish to have trunks in the clump, usually three to five.

You will want most shrubs to have lots of

branches, with some beginning close to the ground. To develop uniform branching close to the ground, start pruning early. Once cuttings are rooted, you have to make a decision about pruning. Pruning tops can delay the development of new roots, so you want to prune only the most vigorous plants when transplanting. If you will be keeping plants in the rooting containers for a while, prune them there. If you want to transplant to other containers as soon as possible, prune a couple of weeks *after* transplanting. New root growth will have started by then. Never prune later than 6 weeks before the expected first frost in fall because pruning alters the natural hormone balance in plants. A plant's internal chemistry must be in balance if it is to go dormant naturally. If the plant does not harden or go dormant, it can easily be injured by average winter cold.

One of the best times to prune to get maximum branching of plants—whether they are in the landscape, transplant bed, or containers—is late winter. For plants that don't consistently respond to pruning, such as mountain laurel, pruning in late winter is a standard nursery technique to obtain a full, bushy plant.

Procedures for transplanting young plants are similar whether the plants are cuttings or seedlings. All the same precautions should be taken. However, you usually have lots of roots and some media with a rooted stem cutting compared to a few roots and no media with a seedling. The stem on a rooted cutting is usually far less delicate than that on a seedling. The connection point between the shoot and a root cutting piece is often delicate the first spring, so care in handling is needed to prevent snapping off the plant you've worked so hard and waited so long to obtain. I'm not suggesting that you should be more careless when transplanting stem cuttings, just noting that you are often dealing with a more substantial plant that is less likely to dry out, snap off, or have its stem crushed in transplanting. *All* cuttings should be treated carefully.

Rooted cuttings should always be transplanted so that they will be growing at the same depth as when they were rooting. Do not plant them deeper or shallower. Cuttings that are planted shallower tend to flop over. When transplanted deeper, their roots may suffocate.

Cuttings should have media or soil firmed around roots tightly enough to hold the cutting erect, but not packed so tight that roots cannot penetrate the media. In addition to hampering root penetration, packing the media too tightly can deprive roots of the oxygen necessary for respiration. Stomping on the ground around newly planted stock may hold the young plants upright but may also kill existing roots and prevent the development of new roots. After planting, gently irrigate the transplants to further settle media around roots and prevent air pockets, but, again, do not stomp.

Two

The Landscape

Plants, like people, live in communities. In much of the eastern United States, most native people and plants lived originally in or at the edge of the forest. A good argument can be made that this is where easterners, including those of us whose ancestors were not native, are happiest. We only thrive, that is, continue to grow and develop beneficially, as long as we are not removed from the forest and forest-edge conditions that once covered our region.

Often we are trying to re-create or enhance a native plant community when we dream of our ultimate landscape. While we must consider the characteristics of each individual landscape plant, even in a very uniform neighborhood of similar plants, we should never forget the community. The community consists of wildlife such as birds and butterflies, humans, and other animals. It is also made up of plants that can take on a seemingly infinite variety of shapes, colors, textures, fragrances, flavors, sizes, and other attributes. All of these characteristics fill an emotional as well as physical need in man. For example, cool shade on a hot day offers both psychological and physical comfort. If a shade tree is on the south or west side of an air-conditioned house, it may answer an economic need in addition to being beautiful by reducing the homeowner's power bill and increasing the real estate value of the home. What value do you place on the entertaining antics of the birds and small animals that are attracted to, sheltered in, and fed by the shrubs and trees in your community? Could you grow beautiful flowering azaleas in the desertlike microclimate created on the south side of many buildings if trees did not cool the earth? None of the plants or other members of the landscape, your community, live alone. They should live in harmony with each other in mutual interdependence.

Putting the pieces of this community back together after it has been altered is one form of landscaping, one too often faced by people who live in homes built on land that was once fields of corn or hay. For these people, planting something to keep the earth from washing away has to be the first priority. Next, unless they can afford to do everything at once, must come the backbone of the landscape, namely, trees and large shrubs. Most of us will add some instant color from herbaceous plants while gradually helping the landscape to develop until we have created something that looks like an open woodland or the forest edge. Even if you are choosing only one new shrub for your yard, remember that it will have to fit into a similar scene. This is the plant community where we feel most comfortable, most secure.

This book is not a text on landscape design. It is about recognizing and growing some terrific native shrubs and trees that can contribute significantly to the community you call your landscape. They grow together with their roots in the same earth, their leaves in the same air and sunlight. For them to share space so that all parts of the community live harmoniously, you'll need to know how to plant, prune, and provide proper conditions for these plants so they coexist attractively in good health. This part of the book should help you learn about these considerations.

Native plants logically seem best suited to a re-created or augmented plant community. Sometimes, however, native *communities* cannot be reintroduced. Native *species* can still be used effectively and will blend harmoniously with the exotic counterpart of one of our native species.

Plants adapt to growing conditions. The climate and growing conditions in some of Asia are very similar to those found from Philadelphia to Florida. As a result, there are often Asian cousins to our native plants. The landscape is yours. As a purist, you may choose only natives. Or you may decide to blend natives and exotics to form the community that pleases you. If you have created a healthy, aesthetically pleasing community, you have succeeded no matter which route you take.

Site Analysis

Look carefully at your yard or garden. Most of us have spots that are dry or wet, shady or sunny, hot or cold. If you live in a swamp, the plants you select for your landscape should be able to tolerate wet feet. Dry or rocky sites might require drought-tolerant plants for a successful landscape. (See Appendixes 3 and 4 for some native plants recommended for moist and dry sites.) If you're living in a thick forest, you're probably going to need to remove some trees and "limb up" a lot more if you want flowering shrubs or small flowering trees to bring interest to your landscape.

Sometimes the hardest thing for gardeners to do is accept the limitations of their gardens. The most successful gardeners recognize these limitations and turn them to advantage by planting appropriate plants. You must match the plant to the site as well as to its place within the community. If the plants you have chosen don't fit, you can change the site by draining it, building a pond, removing trees, etc. If all this change is not an option, and you absolutely must have certain plants that are not adapted to your garden site, choose another place for your garden. I'm a firm believer in making the most of what you've got rather than making major changes in the face of the earth. However, I also advocate adding pools of water and amending soils. To me, this isn't a contradiction. Draining wetlands or leveling mountains pleases me less than building raised beds or diverting driveway runoff water so that it goes around rather than through my vegetable garden. It's a question of the magnitude of change and what makes sense.

Looking around and recognizing what you have is called site analysis. Books are written on the subject. Consultants make a living doing it. All I'm asking you to do is to try to match the needs of your site—whether it's acres or a few square feet—to those of your plants. For example, perhaps you visualize a native dogwood in a moist site. Flowering dogwood (*Cornus florida*) will die if its feet are constantly wet, while silky dogwood (*Cornus amomum*) does perfectly fine with wet feet. The former is a small tree, the latter a large shrub. Both can have white flowers, so if all you want is white flowers, one of these dogwoods should fit your needs. However, if you must have a white-flowering tree in a moist site, neither native dogwood is what you want. Perhaps sweet bay (*Magnolia virginiana*), planted on slightly mounded soil would be a better choice. The habitat description in the plant list should let you know whether a tree or shrub is well suited to the conditions in your garden.

Soil Preparation

Nothing is more important to the establishment and continued health of your plant than proper preparation of the soil prior to planting. To properly prepare the soil you must know your plants as well as some basic soil physics and chemistry.

Soil Physics

The physics of soil—how big soil particles are and how they fit together—often determines whether a plant will die, barely survive, or thrive in a given location. For most of us, it is impractical to make major changes in soil structure. However, you can make certain relatively small changes that significantly increase the chances for chosen species to grow and thrive.

In the poorly drained clay soils so common throughout much of the South, two techniques have proven successful for commercial landscapers so often that they have become standard tricks of the trade. The first of these techniques involves the following steps: turn the soil in your planting bed and rake it smooth; next, borrow topsoil from one area, often an area that will become a path, and pile it onto your freshly turned topsoil. By doing this, you create high and low areas in the garden. Water will accumulate and run off through the low areas while draining away from the higher areas, which now have deeper topsoil. These changes should be subtle, achieved by skimming topsoil rather than digging trenches. Adding 6 inches of soil to a bed can often make the difference between success and failure with plants sensitive to root rot like pieris and many rhododendrons. When building raised planting sites, be careful not to pile soil over the roots of drainage-sensitive, shallow-rooted small trees like flowering dogwood and redbud. However, putting a few inches of soil over the roots of native hollies or tulip poplars may be the only way you can get shrubs established near these greedy feeders.

The second technique for improving drainage is to use organic matter as a soil amendment. My favorite organic matter to use in a situation where I want to improve drainage is southern pine bark, usually from loblolly, longleaf, slash, or other "yellow" pines. Bark from eastern white pine does not work as well. The grade I use is the least expensive, usually sold as mulch. For purposes of soil building, the larger, more expensive pine bark "nuggets" are no better than the cheaper mulch. Spread pine bark to a depth of at least 4 inches over an area at least twice the size of the area to be covered by the branches of the plant you are setting and then mix it thoroughly with the top 6 to 8 inches of soil. Two things will happen: you will increase the air space in the soil, and, since pine bark floats, some of the bark will become a mulch as soon as you water-in your newly planted tree or shrub. Research has shown that the use of pine bark on clay loam soils in the commercial production of mountain laurel resulted in a 50 percent increase in top growth over 3 years when compared to the use of no organic soil amendment.

If you can, use both of these techniques, particularly if you are trying to simulate a mountain or upper piedmont landscape in warmer regions. Using pine bark as a soil amendment as well as mounding soil into raised beds will help to create conditions similar to those of many upland wooded sites, particularly if you use logs or rocks as retaining walls around the beds.

Gardeners who live where soil is sandy may also wish to use pine bark, peat moss, leaf mold, or well-rotted compost as a soil amendment. All will increase the soil's capacity to hold both water and fertilizer as well as helping to prevent erosion. At least 2 inches of organic soil amendment should be used. In most locations, however, 4–6 inches of organic matter should be spread on top of the soil before tilling it in. Unlike mulch, which can be renewed each year, organic soil amendments can practically be applied only at the time of planting. After that, the plant is on its own in coping with soil structure.

Warning: do not use hardwood bark as a soil amendment. Research has shown that it decomposes much more rapidly than pine bark. In the process of breaking down, it may raise the soil pH and cause certain essential nutritional elements to become unavailable to your plants. On the other hand, I've never seen any harm result from using 1–2 inches of hardwood bark as an annual mulch in home landscapes. Apparently, naturally acid soils, rainfall, and some of the fertilizers we use counteract the harmful effects of hardwood bark decomposition when the bark is used as a mulch. Detoxification of the hardwood bark does not take place rapidly enough when it is mixed with the soil as an amendment because roots are forced to grow where the chemical reactions caused by natural decomposition are occurring.

Research on the benefits of mixing organic amendments with the soil in the planting hole has shown mixed results. For many plants, particularly trees, there is no benefit. For others, like the mountain laurel mentioned earlier, the benefits are enormous. In no case, however, have I seen research results that show adding pine bark or finished compost is harmful to an individual species when it is planted in a well-chosen and otherwise prepared site. Since my landscape is a community made up of many different species, I choose to use organic amendments whenever I think they will help a plant become established as a contributing part of that community. I don't really care whether the organic amendment makes the plant grow faster.

If you must have a rule of thumb for the use of soil amendments, try this: use them with woodland shrubs like native azaleas or with plants that naturally grow in moist areas, like buttonbush or sweet bay, but not with plants that grow best where they have good drainage, like flowering dogwood, persimmon, redbud, and New Jersey tea.

Many gardeners want to mix sand with clay soils to improve a planting site. In most situations, when you mix sand with a heavy clay soil, you get a substance similar to adobe, not something you would want in your garden except perhaps for buildings or sculpture. In addition to being heavy, sand won't solve poor soil drainage problems as well as ditches and pine bark. The volume of sand required to make its addition effective is not practical. Spend your money on drainage, pine bark, or water-loving plants. Positive results will occur.

Soil Chemisty

Plants require at least sixteen chemical elements for healthy growth. Six—carbon, hydrogen, oxygen, nitrogen, phosphorous, and potassium—are required in relatively large quantities, so they are called macronutrients or major elements. Fortunately, the first three come free from air and water. Nitrogen, phosphorous, and potassium must be supplied if they are not already present in soils.

Many of the clay and clay loam soils of the eastern United States already have a level of potassium adequate for landscapes, if not for vegetable gardens and annual flower borders. Nitrogen and phosphorous are often lacking in these same soils. The only way to be sure of what you have in your soil is to have the soil tested. Contact your county agricultural extension agent at least a month before you expect to plant your landscape to learn how to have your soil tested, and then get it tested.

How you apply nitrogen and phosphorous to your soil depends upon your own preferences and the local availability of fertilizer materials. Naturally occurring materials high in available nutrients are usually heavy to ship and contain these nutrients in a form that is only very slowly available to the plant. For example, most ground natural rock phosphates can be used, but you must apply a great deal to accomplish the same thing you can accomplish with superphosphate. Ground hard-rock phosphate from eastern North Carolina is an exception and contains 36 percent phosphate. If you can get it, use it like superphosphate. Superphosphate, 0-46-0 fertilizer, is made by treating natural rock phosphate with acids. This makes the phosphorous much more soluble as well as concentrating the amount of phosphorous. You can use either rock phosphate or superphosphate to provide the phosphorous your plant needs.

Generally, with woody plants you want to grow a good, healthy root system while gradually producing a healthy, sturdy stem and leaves. For this reason, I recommend thoroughly mixing your source of phosphorous with the top 6 to 8 inches of planting soil before setting plants. Plants will not rapidly establish a healthy, vigorous root system in soils that are deficient in phosphorous.

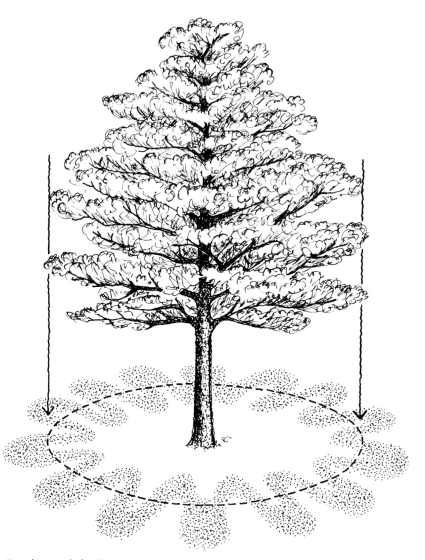

Figure 11. Drip line, with fertilizer applied around it.

Nitrogen, on the other hand, can be applied to the surface of the soil after planting. Use nitrogen-containing fertilizers sparingly to produce gradual, uniform growth rather than lush, often brittle, and unhealthy growth. If you are using a chemical fertilizer, the fertilizer should be applied no sooner than a few weeks after planting. Start spreading fertilizer no closer than 6 inches from the base of the plant and spread it uniformly from there to 6 inches beyond an imaginary line formed by water dripping off the plant's outermost branches (the drip line). The fertilizer should be spread in a circle, with the center of the circle represented by the stem of the plant and the circumference falling 6 inches beyond the drip line (see Figure 11). Larger shrubs and trees should be fertilized in a circle roughly a foot or two inside and outside the drip line. This is where feeder roots are most likely to be found.

If you are using a natural organic fertilizer that is relatively low in nitrogen, such as dehydrated manures or compost, it is hard to apply too much. However, if you are using higher-nitrogen natural fertilizers like blood meal or cottonseed meal, be careful to use them sparingly. It is possible to overfeed or burn newly planted landscapes with these materials. Never fertilize the landscape with fresh animal manures; compost them first.

Three fertilizer elements required in lesser amounts than macronutrients are called secondary elements or secondary nutrients. These are calcium, magnesium, and sulfur. Calcium and magnesium can be applied as ground dolomitic limestone, which is natural rock that is mined and ground very fine so that it will react with the soil. If your soil test calls for limestone, add dolomitic limestone prior to planting and then mix it thoroughly with the topsoil. Take another soil test every three years to see if you need to add dolomitic limestone to the soil surface of your landscape in order to provide calcium and magnesium or raise the soil pH. Sulfur is rarely needed in any eastern U.S. landscape away from the beach. In many areas, all the sulfur a plant could possibly need comes free in the air and acid rain. Sulfur, however, is one of the culprits that may lower soil pH excessively, causing a need for limestone applications to the landscape every few years.

Chemists use pH as a way of determining how acidic or alkaline something is. Soil chemists have determined that most native plants in the Southeast grow best in soils with pH values between 4.8 and 6.5. However, this range is too great for the diversity of plants we choose to put in our landscapes. We are seeking an overall average for our plant community. In seeking this average, I've observed that plants will generally grow well at a pH of one-half unit above and one full unit below that listed as their optimum. Therefore, if you want to grow azaleas, which do best at a pH from 4.8 to 5.3, and flowering dogwoods, which do best at a pH of 6.0 to 6.2, you should be able to have both plants perform well at a soil pH of 5.5 to 5.8. The leaves on the azaleas might not be quite as dark green, and the dogwood probably won't grow new branches quite as long, but both will perform nicely and be healthy.

In most landscapes, using dolomitic limestone before planting to establish a soil pH between 5.5 and 6.0 is a good rule of thumb. Your soil test results should give you guidelines for doing this. When you are planting azaleas, leucothoe, pieris, and other acid-loving plants, you will lower the pH by mixing in organic soil amendments. In places where you are establishing trees that benefit from a bit higher soil pH, mixing a handful or two of ground dolomitic limestone with the planting soil to create a locally higher pH should get the tree off to a good start.

Most soils in the eastern United States are naturally acidic. Our native trees and shrubs have evolved with these soils. As a result, they are well adapted. However, in some locations limestone outcroppings, seashells, or other factors have raised the soil pH to a level at which these plants don't perform well. Certain native plants (see Appendix 6) can tolerate these neutral or slightly basic conditions and should be sought when you are gardening in mildly alkaline soils.

Gardeners in coastal areas or places with limestone outcroppings are faced with a problem opposite from that confronting most of us. Their soil is too sweet or alkaline for many beautiful acid-loving plants. The chemical solution to the problem of soil with an excessively high pH is sulfur. Sometimes this is the only short-term answer. If it is for you, your county extension agent can suggest an appropriate amount of sulfur to apply and techniques for handling this nasty and corrosive, if naturally occurring, powder. However, for the long term—that is, after plants are in the ground and established—continually building the soil with organic mulches on heavily amended soils planted in raised beds may be a better solution. Don't let your azalea roots grow down into soil made of seashells or the plant's leaves will turn yellow from iron deficiency.

The remaining chemical elements necessary for plant growth are needed in tiny quantities. They are collectively called minor elements or micronutrients. If plants are grown at the proper pH, rarely will these need to be applied. Still, boron, copper, iron, manganese, molybdenum, and zinc may occasionally become deficient, particularly in soils with a high or extremely low pH. (Chlorine is a seventh micronutrient, but in my experience working with a wide range of soils, I have never known it to be deficient.) In sweet, or high pH, soils the best long-term solution to a lack of micronutrients is to lower the soil pH with either organic matter or acid-forming fertilizers like ammonium sulfate. On extremely low pH soils, liming with dolomitic limestone should correct the problem. In either case, you are just trying to create a "normal" versus "abnormal" condition for plant growth.

If new leaves on acid-loving plants like azaleas, mountain laurel, and pieris are yellow even though their veins remain green, and you can't wait for them to turn green while the soil pH drops over the period of a few weeks after applying sulfur or ammonium sulfate, you can often turn the leaves green in a few days with a chelated iron spray. These sprays are sold under a number of brand names, including Nu Iron and Sequestrene Fe. Using the sprays will only correct the immediate problem, however, not cure it. Lowering the soil pH and taking measures such as regular mulching with pine needles, pine bark, or other materials that have an acidic reaction when they decompose will help to keep iron deficiency from reoccurring. Rarely do soils lack iron, but even when iron is present in the soil, it can be unavailable to the plant because the soil pH is too high.

Planting

Whether you grow your own woody landscape plants or buy them, they will come to the landscape in two forms. They will either have been field grown in soil or pot grown in essentially soilless media. Despite taking different paths, once these plants reach your garden, they will need to be treated in much the same way.

Most woody plant roots do not grow straight down into the earth. A few tree support roots may anchor themselves by growing deep into the earth, but the great majority of roots are going to grow outward from your shrub or tree. In fact, most of these roots will be in the top 12 to 18 inches of soil. These are the roots that forage for the nutrients and water that sustain plants in the landscape.

For this reason, when you dig a hole to plant shrubs and trees, it is more important to dig it wide than dig it deep. I generally dig the hole a few inches deeper than the rootball and two to three times its circumference (see Figure 12). The reason I go deeper at all is that I'm looking for big rocks, pockets of blue clay, and other obstructions that might make me change my plans about where I want to locate a plant in the landscape. Many more plants die as a result of poor soil drainage and excessively wet roots than die from being too dry, so I try to be aware of potential drainage problems.

If you are planting something that has slightly different nutritional requirements from the majority of plants in your landscape, this is the time to do something about it. If you need a little extra ground limestone or superphosphate, at or before planting is the time to apply it. If you are planting after fall frost, in late winter, or early spring, mixing well-rotted manure or finished compost with soil from the planting hole is an excellent way to get the plant off to a good start. However, be very careful if using any materials that contain much nitrogen, like manure or compost, when planting in early fall. Your plants may enjoy the fertilizer too much, putting on a tender new flush of leaves just in time for a killing frost to turn them black. So, if you are planting a few weeks after the first killing frost, go ahead and use a low-nitrogen fertilizer. But if you are planting in the fall before the first frost, hold off on applying nitrogen until your plant has been exposed to a couple of killing frosts. Any soil amendments and fertilizer should be mixed thoroughly and uniformly with the soil you will be using to backfill the hole.

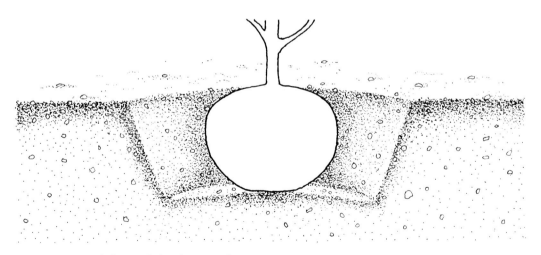

Figure 12. Size of planting hole relative to plant.

When handling your shrub or tree, always hold it by the rootball. This way you support the weight of the plant rather than having the stem and roots carry the weight of media and soil. Many a plant has been ruined because good roots were ripped off with the soil as it fell away from the rootball. Soil is supposed to support the weight of plants, not vice versa.

Mound soil in the bottom of your planting hole and situate the bottom of the rootball on top of the mound as you place your shrub or tree in the hole. The top of the rootball will be even with the level of the surrounding soil or slightly higher than the surrounding soil to allow for settling of soft soil in your planting hole. Never plant a woody plant deeper than it was growing previously. Planting too deep is a major cause of poor establishment and slow growth of landscape plants. In areas with clay or clay loam soils, many gardeners have enjoyed success with rhododendrons and similar root-rot-sensitive plants by using raised beds and organic soil amendments *and* planting so that rootballs appear to be half out of the ground (see Figure 13). The top of the rootball is then covered with about 4 inches of organic mulch, which must be renewed to the original depth of 4 inches each year, usually by raking leaves or pine straw in the fall and top dressing in the spring. The roots of these plants establish themselves in the top inches of soil and the decomposing mulch layer. By planting high in this way, you are mimicking essentially the same soil building that takes place in native rhododendron communities. Wild rhododendron roots are mostly in the decomposed woodland litter and leaf mold, rarely penetrating into the topsoil more than a few inches. With this technique, you must be careful, however, to water regularly the first year and during prolonged periods of dry weather thereafter. Shrubs planted in this way tend to be far more likely to dry

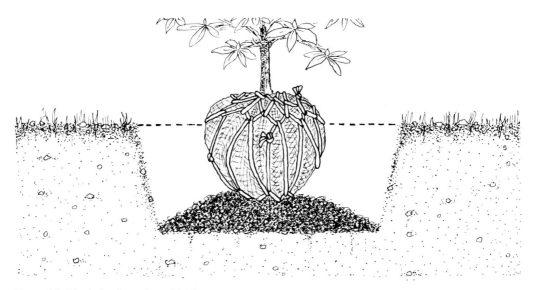

Figure 13. Rhododendron planted high.

out. You are protecting against conditions becoming too wet, so you must be willing on occasion to tolerate their being a bit too dry and be prepared to react with a sprinkler.

Plants that come balled and burlapped should have the root covering removed. If you are absolutely sure that the covering is natural fiber, it may be removed from the top and sides of the rootball after the plant has been set in place and then left in the hole, where it will rot after being covered with backfill soil (see Figure 14). You should be aware, however, that burlap made entirely of natural fibers has become uncommon in some parts of the United States. Most burlap contains synthetic fibers that will not decompose, so the safest course of action is to remove the fabric covering altogether.

Likewise, plants grown in containers should be removed from their containers before planting. (To most of you this may seem obvious, but you would be surprised how often I have been called into private or commercial landscapes where plants are not growing and find plants with their roots in plastic nursery containers buried in the soil. It's amazing—and a testimony to how tough our landscape plants are that they live as long as they do when planted this way.) The best way to remove smaller plants from nursery containers is to turn them upside down, with the stem between your fingers or hands. Support the weight of the rootball as you gently, and sometimes not too gently, tap the edge of the container against something firm like a rock, garden cart, or pick-up truck.

Once the covering has been removed from the rootball, set the plant on the mound in the hole and inspect the rootball. If the top of the rootball is perfectly flat, look to see if there are any roots in the topmost soil or media. If not, remove the soil or media until you get to roots, which will

Figure 14. Natural-fiber root covering removed and left in hole to decompose after planting.

Figure 15. Handling a slightly root-bound plant.

often create a rootball with rounded shoulders. Ideally, the roots of container-grown plants will have grown just through the media, reached the edge of the pot, and started to grow around the edge of the rootball when you are ready to plant. If the sides and base of a container-grown rootball are an impenetrable mass of roots, gently tease the roots away from the sides of the ball. If this is difficult because the roots are impossibly intertwined (the plant was "rootbound" in the pot), cut into the rootball with a knife or pruners and pull the cut roots away from the sides. (See Figures 15 and 16.) Planting this fuzzy, damaged-looking rootball instead of the solid clump that emerged from the pot will increase your chances for long-term success.

Keeping the plant vertical, backfill around the rootball so that the new soil will hold the plant up. Fill the hole at least three-quarters full (a 12-inch-deep hole, for example, should be filled to within 3 inches of the top); then fill the remaining space at the top of the hole with water and go get some refreshments. This will allow the water to seep into the soil, filling any air pockets, and keep you from meddling with the process. Fill the hole with water one more time, letting the water drain into the soil, and then backfill the hole so that it is level with the surrounding soil.

You should bear in mind that it is impossible for the soil to be exactly level. You have added air spaces, a plant rootball, and possibly some soil amendments. Use the excess backfill to make a low containment dam or dike around the edge of your planting hole and then gently sprinkle the area with water to firm the topsoil and bind everything together. Do not stomp on the soil to firm it. You want tiny new roots to become established in this soil.

Once the last water has seeped into the soil, mulch. Many things can be used for this purpose, but my mulch of choice is pine bark or pine needles. For a small job like planting an individual shrub or tree, a couple of bags of pine bark mulch or a bale of pine needles is convenient and effective. (See Figure 17.)

Mulch

Readers should be warned about some of an author's biases. There are mulchers and non-mulchers among horticulturists. I am a mulcher.

Mulch is a layer of organic or inorganic material spread on the earth to modify rapid changes in environmental conditions. Mulch acts like a blanket to keep the earth warm in cold weather and cool in warm or hot weather. A mulch will also help to prevent evaporation of soil moisture, reducing or eliminating the need to water between rain showers. It can also help to reduce splashing of soil during heavy rainfall, reduce erosion, particularly in newly planted landscapes, and nearly eliminate the baking and cracking of clay soils during a drought. While I have seen very effective, and occasionally attractive, mulches of pebbles, calcined clay, crushed brick, and other inorganic materials, I prefer organic mulches for their natural-looking appearance.

Figure 16. Handling a severely root-bound plant.

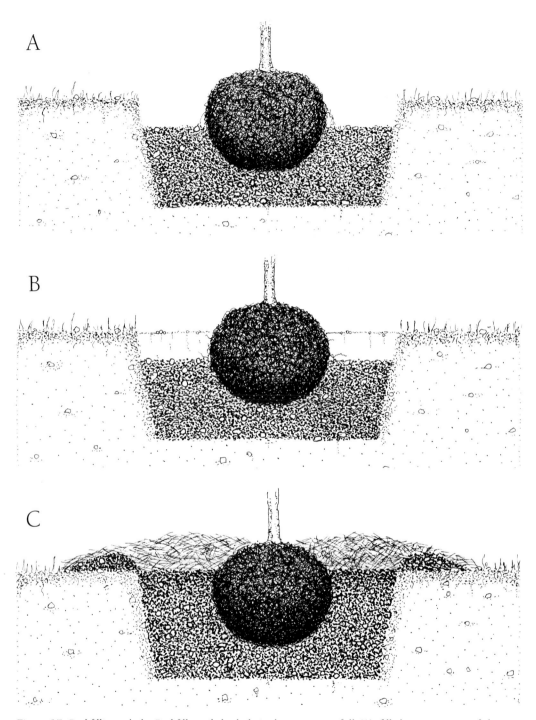

Figure 17. Backfilling a hole: Backfill until the hole is three-quarters full (A); fill the top quarter of the hole with water and let it seep down into the backfill (B); after watering and allowing the water to seep down once more, finish backfilling the hole until it is level with the surrounding soil, create a low containment dam, and mulch (C).

Organic mulches are a bit more work than inorganic mulches, but not that much more. Weeds grow in both and must be removed from both. Both need to be replenished as they settle into the topsoil. But an organic mulch, when managed properly, slowly improves the topsoil as it decomposes. Leaf mulches are the most rapid to decompose, thank goodness, or we would soon be buried in leaves. Pine needles are next, followed by wood chips, hardwood bark, and pine bark. I've used all kinds of things as a mulch, including rice hulls and coarse sawdust, but prefer leaves, pine needles, or pine bark when I can get them. I don't like most peat and sawdust for use as a mulch because it will bind together, taking on a cardboardlike texture when it dries. Also, the soil microorganisms that decompose sawdust require nitrogen. These microbes are more efficient at getting the nitrogen than plants are, so they rob fertilizer, turning plants yellow during the first stages of sawdust decomposition. In addition, sawdust from a few trees (like black walnut) are toxic to some plants. In most cases, pine needles are my favorite mulch because they don't float or blow away and they won't mat together to shed water or prevent the emergence of spring bulbs and herbaceous plants I'd forgotten about.

When working with organic mulches, it is important to avoid what Dr. Francis Gouin calls "rich man's disease." This malady manifests itself when you apply 6 inches of mulch this fall and then tell the gardener to apply 6 inches next spring and again next fall. You are adding mulch far faster than it can decompose, so the roots of your plants are getting farther and farther from the surface and well-aerated soil. As the mulch does decompose, it forms a thatch layer of a peatlike material that contains little plant nutrition but is often the only place where the soil is adequately aerated, so roots must grow there or the plant will die. As a result, a few years after planting, your shrubs and trees look like they are starving, with progressively smaller leaves. You can increase your fertilizer rates, but managing the system you have created is nearly impossible because this is also a system that will dry out very rapidly, making it necessary to water more often. Watering a soil that doesn't contain much air space creates conditions that favor root rot.

Don't get into this cycle. In much of the East and South, a 4-inch layer of mulch is all you need. On hotter, sandier soils, 6 inches may be needed. As the mulch decomposes, renew it. Six inches of leaves in the fall will probably be 2 inches in the spring due to normal decomposition. Renew the mulch after you check its depth. Pine bark and longleaf pine needle mulches are slower to decompose than other types, but they will also need renewing annually.

Competition

"Spatial relationships" is a favorite topic for landscape architects and designers. They all seem to agree that there is no such thing as empty space but commonly refer to places where there are no plants, rocks, or buildings obstructing the view as empty space. Ground-hugging plants like mowed turf or ground covers don't seem to count as something that fills empty space. When you landscape, you either fill or create empty space.

The concern of garden designers is to make things look good. Some plants look good crowded together like a formal hedge, while others seem best when set off by themselves as plant statues, with lots of open space around. In nature, often our most permanent memories of plants are those of specimens that are set off by themselves at the top of a hill or leaning over water.

Spatial concerns can be translated, in practical terms, to the question, How far apart should plants be? The tendency is to plant too close together when plants are small, so that they crowd each other a few years later. If you are willing to remove some plants when they get too crowded, go ahead and plant close together. It's your garden. However, in terms of planting for the health of the plant, please use some ideas that I hope will seem like common sense. Most plants need space of their own when they are becoming established. Once roots are well established, plants are better able to battle for available sunlight and air.

For gardeners who want rules, here they are. Large trees should be planted no closer together than 50 feet apart or 25 feet from a building. These trees are the forest giants like many native maples, oaks, and pines, and tulip tree. Smaller trees should be planted no closer together than 25 feet apart or 15 feet from a building. This group includes dogwoods and redbuds. (See Appendix 2 for size classifications of the native trees and shrubs discussed in this book.) If the ultimate height of a tree, according to a reference book, is under 50 feet, you can generally consider it a small tree. Large shrubs planted 10 feet apart will eventually have their branches touch by growing 5 feet on either side of the planting holes. Small shrubs planted 6 feet apart will usually touch 5 years after planting. Nature does not tolerate a vacuum. Given a chance, most plants will fill the space you allot to them and a little more.

I don't use these rules. I'm happy if a plant will have a few years of glory filling a particular space. If I fall in love with a plant, I remove or prune back its neighbors. If I love its neighbors as well, I prune according to what seems appropriate. There is a place in my back yard where the border goes from mostly sunny to mostly shade in a few feet. A white-berried beautyberry, *Callicarpa dichotoma*, grows next to *Itea virginica* 'Henry's Garnet'. By October each year the beautyberry is about to overwhelm the itea. Frost hits, beautyberry leaves are blackened, and sometime during the next months I severely prune the beautyberry. Flowers and fruit on the fast-growing beautyberry are on the current season's growth, so the void created by the heavy pruning will rapidly be filled during the next growing season. Meanwhile, the semievergreen 'Henry's Garnet' has enough light and its place in the border until the beautyberry threatens again late in the next growing season. Annual pruning is required as long as this combination strikes my fancy. These two plants are otherwise maintenance free, so this seems little enough work to keep two of my border favorites performing in a way that pleases me.

The common sense part of spatial relationships comes in plant selection and placement. For example, large, fast-growing plants are likely to overwhelm slow-growing, small plants. Nothing seems to grow well in the dense shade beneath a southern magnolia (*Magnolia grandiflora*), so why plant anything there? Perhaps that's the place for a bench where gardeners and guests can relax on a warm summer evening, watch fireflies, and be enveloped by the delicious fragrance of the queen of our native flowering trees.

Sometimes a change in plant selection as simple as choosing a different species is all that's needed to solve a spatial problem. Eastern hemlock, *Tsuga canadensis*, has a relatively shallow, very aggressive root system. Any shrub planted to take advantage of the lush green visual background provided by eastern hemlock is going to need extra attention because the hemlock's roots are so aggressive. If a gardener chooses Carolina hemlock, *Tsuga carolina*, instead, competition will be greatly reduced while the visual effect will be much the same. Why? Carolina hemlock is a "tap-rooted" plant: most of its roots will grow down rather than out (see Figure 19). Carolina hemlock is also slower growing for about the first ten years of its existence, however, and therefore less popular than eastern hemlock in the nursery trade, where most customers buy on the basis of growth and price because they don't know one hemlock from another. Nurseries can sell a 5-foot-tall eastern hemlock for less than a 5-foot-tall Carolina hemlock because they have had to grow it for at least two years less. Since most buyers aren't willing to pay more for a Carolina hemlock, few are found in garden centers.

Choices like this abound in the garden. Folks who live in the eastern United States

Figure 18. The dense shade of southern magnolia provides a poor site for planting but an excellent setting for a restful bench.

are blessed with an enormous variety of native plants that are the envy of gardeners throughout the world. Let's use them!

These choices don't need to be overwhelming. The average landscape can support only so many shrubs and trees. Make a list of the ones you like, read about them, find out if they are available, and then make your choices. If you choose to plant small trees, you probably can't fit in more than a couple. If you are looking for something to put in a site that is well drained, flowering dogwood and redbud should be easy to find. Sourwood or sweet bay may require a more diligent search.

One aspect of plant competition is not a question of spatial relationships and may not even fit in the category of common sense. Grass is often the leading pest for nursery growers, worse than insects, diseases, and weeds. The types of grass that have proven most devastating to woody plant growth are the cool-season perennial grasses like bluegrass and tall fescue. These are also the grasses of choice for lawns throughout most of the eastern United States.

Most of us like the vision of a green carpet leading up to our shrub borders or lying under our shade trees. Once woody plants are well established, if you want to try to create a lawn right up to the base of these plants, go ahead. But for the first couple of years a plant is in the ground, please

Figure 19. Comparison of eastern hemlock and Carolina hemlock root systems.

mulch over the roots of the plant rather than trying to grow grass above them. For those first few years, consider grass a weed. Grass uses water and fertilizer intended for your woody plants just as most other weeds do. In fact, it is often much more efficient than other weeds. Some grasses, like tall fescue, have also been shown to exude chemicals that are actually toxic to other plants through a phenomenon known as allelopathy. Once woody plants have established their territory, usually after about three years, most are capable of competing with grass. The grass doesn't grow as fast as it would if the woody plants were not there and vice versa. Some years the grass does better, and some years the woody plants do better, but both survive except in instances where woody plants cast enough shade to cause the grass to die out.

For reasons related to ease of maintenance and lack of damage to woody plants, I prefer for no grass to grow within about 10 feet of the base of dogwoods, redbuds, and other small trees. If you don't need to mow or use a string trimmer near the base of these trees, you are less likely to damage the trunks. This physical damage, when inflicted, often invites disease or insect infestations.

Fertilization

Teenagers can consume great quantities of food and remain healthy, while most older adults on the same diet would become obese and unhealthy. Much the same thing is true of plants.

When plants are being grown in a nursery and shortly after they go into the landscape, we generally want to take advantage of juvenile growth. Younger plants can benefit from extra nutrition to support rapid growth; however, after plants become established, far lower levels of nutrition are needed, with only periodic renewal of certain nutrients. Plants that are given high levels of fertilizer once the landscape is established may have greater pest problems and will need far more pruning to maintain balance within the landscape community. For this reason, I suggest a separate fertilizer program for individual plants that are just getting established, usually for the first year or two in the landscape. Then a maintenance-level program can be used for plants that seem established. (See Appendix 1 for recommended growth and maintenance fertilization rates.)

These fertilizer programs will need to be adjusted to meet the needs of each garden. By spending time in the landscape, you will develop a feel for when a plant seems a bit hungry. Leaves may be smaller or the color just not right on hungry plants. Use the suggested fertilizer maintenance program throughout the landscape; then add a little extra fertilizer for new plants or for ones you have discovered are a bit hungry. For plants that are just getting established in the landscape, this "little extra" is twice as much as the maintenance rate. Although not a very precise method, the way I do this is to apply fertilizer at the maintenance rate and then go back and apply the same amount again in the area where the feeder roots of the new plant are becoming established. Rarely do I actually measure the amount when I return to give plants a boost. I can see how much is already there, so I just add about the same amount again.

Most woody plants absorb fertilizer nutrients with greatest efficiency in the fall, late winter, and early spring, just before new leaf and shoot growth starts. Therefore, it is important to apply fertilizer ahead of these periods of above-ground growth, when the plants can most effectively use the fertilizer. It should be there when new root growth takes place. Timing is complicated by the fact that applying too much fertilizer in the fall can keep a woody plant from going dormant. Lack of proper dormancy may lead to injury from cold that the plant could normally withstand. Therefore, the amount of fertilizer that is applied is just as important as when it is applied.

In Zones 6 and 7, fertilizer can safely be applied to woody plants in mid-October at the same time the lawn is fertilized. I like to apply about one-third of a plant's normal requirement, broadcast over the total root feeding area, in the fall, with the remaining two-thirds applied in mid- to late March. In warmer areas, which include Zone 8 and the cooler portions of Zone 9, wait until early November for the fall fertilizer application. Get the Zone 8 spring application in place while it is still winter, generally about Valentine's Day, February 14. In Zone 9, put out your "spring" landscape fertilizer within a few days of Ground Hog's Day, February 2.

How much fertilizer is enough for your plant community? This obviously is going to depend upon your soils as well as the members of your garden community. I've found a good starting point for a fertility maintenance program is 1 pound of actual nitrogen per 1,000 square feet of land, applied in a balanced fertilizer. A balanced fertilizer is one that contains nitrogen, phosphorous, and potassium, like 6-6-6 or 10-10-10. The numbers, however, do not have to be balanced. In fact, for most plants that are chosen for flowers and fruit, I prefer the balance to be in favor of the last two numbers (i.e., phosphorous and potassium), such as 5-10-10.

To calculate how much fertilizer you actually need, think in terms of both the area to be covered and the analysis of the fertilizer to be used. Next, consult Appendix 1, where I've given volume measures for different areas. I used to say that a 1-pound coffee can full of 10-10-10 contained 2 pounds of 10-10-10. Half a can was enough for 100 square feet. This still works, but fewer people are drinking coffee that comes in cans these days, so I've converted to cups. These are not precision measures, so once you know about how much to apply, wash the cups and return them to the kitchen.

I do my area calculations based on geometry and the length of a step or my boot. My boot is about 1 foot long. My step is about 3 feet long. Therefore, if I have a roughly rectangular bed 2 steps deep and 11 steps long, it is 6 feet (2 steps × 3 feet) deep and 33 feet (11 steps × 3 feet) long. Since the area of a rectangle is determined by multiplying its length by its width, the area of my bed is 198 (33 × 6), or about 200, square feet. These measurements can all be done with a tape measure and a map to record the size of each area to be fertilized if it pleases you. Steps suit me better.

Using this same example, if I wanted to fertilize with 5-10-10, I need to remember that application rates are based on 1 pound of nitrogen per 1,000 square feet and that the numbers on the fertilizer bag are expressed as percentages or decimals. In other words, 5-10-10 contains 5 percent nitrogen; the fertilizer is .05 nitrogen. An easy way to remember this is to think about 100-pound bags of fertilizer. Each 100-pound bag of 5-10-10 fertilizer contains 5 pounds of nitrogen. Since you only want 1 pound of nitrogen per 1,000 square feet, you want $1/5$ (20 percent, or .2) of 100 pounds of 5-10-10. If you multiply .2 times 100 pounds, you come out with the 20 pounds of 5-10-10 fertilizer that is needed to provide 1 pound of nitrogen per 1,000 square feet.

In the bed measured for this example, there were 200 square feet rather than 1,000 square feet, and 200 divided by 1,000 is $1/5$ (20 percent, or .2). Multiplying .2 times 20 pounds of 5-10-10 gives you 4 pounds of 5-10-10 fertilizer as the total needed on the plant bed. (For those with less need for explanations, divide the area, .2 [of 1,000 square feet], by the nitrogen analysis, .05. The answer will still be 4 pounds.) One-third of this should be applied in the fall and two-thirds in the spring. One-third multiplied by 4 pounds equals about $1 1/3$ pounds to apply to the bed uniformly in the fall, with the remaining $2 2/3$ to be applied in the spring.

These calculations may seem confusing. They are relatively simple if you keep in mind that you are trying to figure weight (pounds) of fertilizer to apply to an area (square feet). These calculations are usually mastered in elementary school, so feel free to ask your children and grandchildren for help. Then write down their answers so you don't have to repeat the calculations next year when they may not be around. The other thing to keep in mind is that these are not precise numbers. A *little* more or a *little* less fertilizer is not going to make an enormous difference.

How do you figure for trees in lawns or when you want to use organic fertilizers? Trees in lawns should be fed with the lawn. In most cases, that is all they will need. Organic fertilizers have an analysis printed on the bag if you buy them at a garden supply store. Composted cow manure may be only .5-.5-.5, but it does have an analysis. Since it is only 10 percent as concentrated as 5-10-10, you will need to use ten times as much composted cow manure to provide plants with the same amount of fertilizer they would get from 5-10-10. That's right, you'll need 40 pounds for that 200-square-foot bed. Cottonseed meal, however, may have an analysis like 6-1-1, so you will need to use less of it, because it contains 120 percent ($6/5$) of the nitrogen in 5-10-10. Use the same calculations for both man-made and natural fertilizers.

When fertilizing with homemade compost, I generally save the compost for the cut flower or vegetable garden because there is never enough to go around. However, assume your compost contains 1 percent nitrogen when calculating how much to use. This figure will probably never be accurate, but it should be close enough. More landscapes are harmed by too much fertilizer than by too little. If you have enough homemade compost to cover all the borders and beds, use it. If you think there is too much, save it or share it with your neighbors. You'll quickly make friends with those lucky enough to get your brown gold.

Questions about using wood ashes in the garden frequently arise. The nutrient content of wood ashes is variable depending upon the species of trees burned and how they have been handled. Nutrients in wood ashes are very water soluble, so ashes should be kept dry until they are spread. Wood ashes, like lime, are alkaline and should be applied only to acid soils; they should never be applied to acid-loving plants like mountain laurel and rhododendrons. I use wood ashes on the lawn and vegetable garden, never in the shrub border. I figure it is better to spread wood ashes than to send them to the landfill, but I don't count on them for much fertilizer. A report from Virginia Tech suggests that you should regard dry wood ashes as a 0-2-5 fertilizer and apply them at a rate of 10 pounds (two 10-quart pails nearly full) per 1,000 square feet per year.

Many of these calculations in regard to fertilization are already done for you in Appendix 1. You will have to figure out how much area you need to cover on your own, but after that you can refer to the table for volume measurements. Working from the calculations presented in Appendix 1 will also create a use for empty plastic milk jugs or old bleach bottles that originally contained *1 gallon* of fluid. Cut the top out with a sharp knife or hacksaw at about the level to which the bottle or jug was filled with liquid. The resulting container becomes a handy measuring device. For example, this container, when not quite half full of dolomitic limestone, contains about the same amount as you would need to apply to 100 square feet of bed if your soil test report called for 1 ton of limestone per acre. Totally full, it contains the amount of 10-10-10 you need to apply to 1,000 square feet of landscape before buds break in the spring.

Pruning

Native woody plants usually respond to pruning as well as exotics do. If your landscape calls for a well-branched, compact native plant, and your plant needs pruning, prune it. You would prune an exotic or a hybrid under the same circumstances. Native plants are still plants and respond as plants to all sorts of cultural practices.

Only two reasons exist for pruning. You are pruning either for the health of the plant or for the health of the landscape. Whole texts and many hours of lectures are devoted to all the possible pruning situations that might face you, yet they often fail to mention exactly your problem.

Pruning for the health of the plant includes removing crossing branches that might be rubbing against each other as well as removing diseased or damaged tissue. You may also need to remove overhanging branches (this process is usually referred to as "limbing up" by arborists) from another plant to permit sunlight to reach a plant or permit greater air movement to lessen chances of disease. I consider this type of pruning to be for the health of the plant and feel comfortable pruning at any time of year except within 6 weeks of the anticipated first-frost date in the fall. Pruning too late in the growing season may stimulate new growth that will be damaged by frost.

Pruning for the health of the landscape is very personal. For example, I am over 6 feet tall, so any branches under 6 feet growing where I have to mow or walk regularly are removed from my home landscape. This certainly isn't for the health of the plant. It is for my convenience and the protection of my head. I consider this to be pruning for the health of the landscape. Overhanging plant growth that rubs on the house, makes passage down a path difficult, or obstructs a view all fits into this category. I try to anticipate this type of growth and prune accordingly. This type of pruning is done during the winter. I plan to do it on a warm, sunny January day but usually end up pruning on a cold, damp day in early March.

Native trees and shrubs have a bad landscape reputation in some circles because the plants have not been treated properly. Part of this proper treatment is pruning. If you want low-branched, bushy mountain laurels and deciduous azaleas, prune them hard in late winter. If you prune many shrubs in late winter you'll get multiple breaks that translate into multiple branches. These same plants pruned in mid-summer will often only reward the gardener with one or two new shoots that result in a plant nearly as leggy as it was before. For plants that have flowers only at the tip of branches, pruning at the right time will mean more flowers. However, if you have a spring-flowering plant, you will be pruning off the flowers with winter pruning.

Therefore, if a plant needs to be reduced in size but is sufficiently well branched, prune it soon after it finishes flowering or prune only part of the plant each year. For plants like mountain laurels and azaleas, if you prune the first year or two after they go into the landscape, before they are mature enough to produce many flowers, you can establish the form of the plant while it is young and rarely need to prune as the plant matures. The saying "as the twig is bent (pruned), so grows the tree" may be used in child development lectures more than horticulture these days, but its original use was in horticulture. Too little rather than too much pruning is done in most gardens.

The reputation of native plants also suffers from the belief that they flower only every second or third year. There are a few reasons for this blooming cycle that the gardener can circumvent, enabling flowers to appear every year. To promote annual blossoms, take the following precautions. (1) Water during a drought. (2) Fertilize, but don't overfertilize. (3) Make sure that plants get enough light year after year. This will mean pruning back overhanging and encroaching branches every few years. (4) If there is a particularly heavy bloom, remove the resulting seeds from shrubs and trees. With herbaceous perennials, this is called "dead-heading." This form of pruning is for the health of the landscape (so you'll get flowers the next year) since the shrub or tree would not die from producing a large crop of seeds. Generally dead-heading is more important on shrubs than trees. If you need some seeds to propagate the tree or shrub, feel free to leave enough seeds to supply your needs, but remove the rest so the shrub or tree's resources can be spent on healthy growth and forming flower buds for next year rather than ripening lots of seeds.

Part Two

Plant List

The plants listed below are those among the many native woody plants of the eastern United States that I think are most likely to appeal to amateur gardeners. Their primary appeal is their showiness: their impressive display of fruit, flowers, or foliage or their ability to attract the attention of wildlife as well as gardeners.

Although the list provides specific information on propagation techniques and landscape requirements as necessary for individual plants, no attempt is made to recapitulate here the general information on the propagation and cultivation of natives that appears in Part 1. For general suggestions read the sections on the form of propagation (seeds, root cuttings, or stem cuttings) that interests you; then supplement those instructions with whatever specific advice is found in the plant list.

The plants are arranged in alphabetical order, according to their scientific names. (If you want to look up a plant but don't know its scientific name, check the index of common names at the back of the book.) The scientific name of each is followed by its common name (or names), an indication of its habitat, and some remarks on propagation. Comments on landscape qualities (size, color of flowers or foliage, etc.), considerations for cultivation, and other matters that seem worthwhile (folklore related to the plant, medicinal or culinary uses for the plant, origins of plant names, and so forth) are also included.

In a number of instances, more than one species from a particular genus is included in the plant list. For example, both wild and oak-leaf hydrangea (*Hydrangea arborescens* and *H. quercifolia*) are worthy of comment, as are three different dogwoods (*Cornus*) and four hollies (*Ilex*). In such instances, general comments are either grouped together following the last-listed species (as with the hydrangeas and hollies) or offered species-by-species (as with the dogwoods). Five genera—*Magnolia, Prunus, Rhododendron, Rosa,* and *Viburnum*—call for general comments on the genus as well as comments about each of the species listed. Those comments have been provided under the genus name before the listing of individual species. The number of *Rhododendron* species worthy of notice are so numerous that the listing and comments for them are divided into two parts, one for evergreen rhododendrons and one for deciduous azaleas.

Acer leucoderme

Chalk maple, southern sugar maple

Habitat: Well-drained to rocky woods, Florida to Virginia. Zones 5–8.
Propagation: Seeds are shed in late summer to early fall, often before expected. Germination requires 2–3 months cool stratification. A high percentage of seeds on most trees aren't capable of sprouting, so you must plant lots of seeds to get only a few plants.

Many maples can be rooted under mist or in a poly tent using late softwood or early semihardwood cuttings treated with .5–.8 percent IBA powder. However, percentages are often low. Rooted cuttings should not be disturbed until growth begins the following spring.

Figure 20. *Acer leucoderme*.

This is the maple with brilliant foliage in all shades from yellow and orange to brilliant and deep red that you see on early autumn drives through the southern piedmont. For transplanted Yankees who must have sugar maple color each fall, choose chalk maple. The leaves are smaller than those of northern sugar maple, and the tree itself is smaller too, rarely taller than 25 feet; but the colors are just as brilliant. Fall foliage color arrives early and, with cooperation from Mother Nature, lasts.

The "chalk" refers to the light-colored bark, not a need for lime in the soil. Chalk maple will grow well in slightly acid to acid soils, in well-drained to dry soils, and in full sun or partial shade. The problem, to date, has been finding this tree in nurseries. The time when it is most noticed, in foliar brilliance, is a month or more after seeds have been shed. As a result, little seed is collected or sown, yielding very few trees available in nurseries.

Acer pennsylvanicum
Striped maple, moosewood, goosefoot maple, whistlewood

Habitat: Striped maple is an understory tree or tall shrub, except near mountain balds. It

is most often found in moist, shaded woodlands, but I have seen it growing on rocky hillsides where little else can survive. Georgia to New England. Zones 3–7.
Propagation: Seed germination requires 3–4 months cool stratification. Male and female trees exist, but only females have seeds. Treat cuttings the same as for *A. leucoderme*.

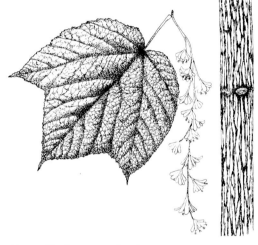

Figure 21. *Acer pennsylvanicum*.

The most distinguishing landscape characteristic of this large shrub or small tree isn't its interesting green flowers, which, unlike those of other native maples, are terminal and pendulous. Instead, this shrubby maple is known for its consistent, bright yellow fall foliage color.

At the edge of a woodland, striped maple will seem lit from within during fall foliage season. Keep it there. Striped maple does not compete well with grass and doesn't respond well to fertilizer. In a natural setting, however, its bright color is as welcome as the green and white stripes on the brown trunk are interesting.

Aesculus octandra
Yellow buckeye

Habitat: Moist woodlands along the Appalachian Mountains from Georgia to Pennsylvania, occasionally in rich upper piedmont coves. Zones 3–8.
Propagation: Collect seeds as soon as capsules start to split in the fall; otherwise, squirrels will enjoy the rich, mahogany brown seeds, leaving you the husks. Do not allow seeds to become dry. Fresh seed is essential. Stratify seeds for 4 months or plant them in the fall 2 inches deep and allow to naturally stratify during the winter.

There are at least seven different buckeyes native to the region of the eastern United States shown in Map 1, plus others introduced from further west or overseas. I've chosen the three I feel are best for the landscape. Most buckeyes will have a five-leaflet palmate leaf and upright, distinctive flower panicles.

Yellow buckeye will grow to become an 80-foot tree with creamy yellow flowers in late spring, dark green disease-resistant foliage all summer, and yellow-orange leaves in the fall. For people and squirrels, the real fall treasure is the buckeye nut, borne mostly in pairs within a leathery husk. The rich, glossy brown nuts are said to bring good luck to those willing to carry one in their pocket. The soft, easily worked wood is a favorite of mountain carvers.

If you don't have moist, rich, well-drained soil, plant something else. To perform best, yellow buckeye needs the right soil.

Aesculus parviflora
Bottlebrush buckeye

Habitat: Low, often moist woodlands just above the flood plain in the coastal plain and piedmont, Florida to South Carolina. Zones 4–8. Rare in the wild.

Propagation: Seeds should be planted as soon as possible after they are collected and never allowed to dry out. Some minimal cool stratification, for perhaps 30 days, seems to be beneficial.

Softwood cuttings from certain plants will root in only fair percentages under mist or a plastic tent. The only way to know whether you have a plant with the potential for its cuttings to root well is to stick some cuttings. Rooting is enhanced by treating with .3 percent IBA powder or 1,000 ppm IBA liquid. Higher levels of IBA seem to be toxic. The preferred method for propagating bottlebrush buckeye commercially is root cuttings.

Perhaps the most striking buckeye for landscape use, this is a large shrub that becomes a focal point regardless of where it is placed in the landscape. Imagine a rounded shrub 10 feet tall and half again as wide with bold buckeye leaves. Now top it with a cloud of creamy white, upright flower spikes inviting the dance of butterflies and hummingbirds in July when shrub bloom becomes precious. Whether placed in full sun as a specimen or at woodland's edge, bottlebrush buckeye is a traffic stopper.

Even though you are not likely to see bottlebrush buckeye growing in the wild, I include it because it is simply too good to be missed in the large landscape. Thanks to work currently underway in the Philadelphia area, improved bottlebrush buckeyes should appear in better garden centers and mail-order catalogs during the next few years.

The reason more plants have not been available from nurseries is that plants sold produce few, if any, viable seeds. Therefore, much more tedious methods of increase such as softwood cuttings, root cuttings, and division must be employed.

Why aren't more viable seeds produced? (1) Most flowers are male. (2) A long, warm growing season seems to be needed to develop viable seeds. (3) It has been speculated that nearly all plants currently sold in the nursery trade came from one source originally. In natural populations viable seeds are produced, so superior selections have been made and planted near current bottlebrush buckeyes. If no special pollinating insect must be imported as well, enterprising nurseries hope to be overwhelmed by large crops of viable seeds that should lead to plenty of bottlebrush buckeyes for everyone.

Regardless of where your bottlebrush buckeye originated, plant it where it will receive at least a few hours of direct sunlight and an ample supply of moisture. For best results, limed, moderately fertile soils should be chosen. However, sunshine and moisture are more important than fertilizer and lime with this stunning shrub.

Aesculus pavia
Red buckeye

Habitat: Moist pinewoods, bottomlands, and occasional wooded bluffs, from Florida north to Virginia. Rarely found growing wild above 1,500 feet in elevation despite being hardy in Zones 4 to 8.

Propagation: Seeds must not be allowed to dry out. Planted fresh into warm soil, they tend to germinate almost immediately and then get zapped by frost. To avoid this, I usually plant them in the shade, water, mulch, and hope for the best—spring rather than fall germination.

I've never tried rooting red buckeye cuttings. If you wish to try some, follow the suggestions for *A. parviflora*.

Figure 22. *Aesculus pavia*.

Red buckeye is a small tree or shrub. It is used effectively in many colonial restorations and, surprisingly, as a cut flower.

A. pavia is much more tolerant of less-than-perfect soils than are *A. octandra* and *parviflora*. In moist, well-drained soils it may reach 30 feet in height, but most plants are large shrubs, 10–15 feet tall and nearly as wide. This buckeye is a good specimen plant, which is also suitable for the woodland edge. It will flower in partial shade but often suffers leaf scorch in a sunny southwestern exposure.

Red buckeye's flowers are its most prominent feature, with individual flowers an inch or more wide, medium red in color, and borne in panicles up to 6 inches long and 3 inches wide. Specimens in bloom almost always stop strollers, attract hummingbirds, and inspire comment. The selection 'Atrosanguinea' has flowers that are deeper red in color.

Amelanchier arborea
Serviceberry, sarvice, shadblow, Juneberry

Habitat: Two varieties of this tree are commonly seen. *A. arborea var. arborea* can be found in upland woods throughout the piedmont as well as into the mountains and upper coastal plain from Florida to Newfoundland. *A. arborea var. laevis* is found in rocky woods and the upper mountains from Georgia north to Newfoundland. Zones 4–9.

Propagation: If you can beat the birds to the fruit in June, seed can easily be removed by rubbing the fruit against a fine screen and washing away the pulp. Seeds require 3–4 months' cold stratification.

Rooting softwood cuttings using .3 percent IBA powder is supposed to be possible, but I've never succeeded at it. 'Prince Charles', a *laevis* selection, grew vigorously for me from tissue-cultured microcuttings. Future selections will probably be available

Figure 23. *Amelanchier arborea*.

via tissue culture propagation, a modern laboratory procedure in which a whole plant, exactly like the parent, can be grown from only a few carefully chosen cells.

See the comments following *A. canadensis* below.

Amelanchier canadensis
Serviceberry, sarvice, shadbush, Juneberry

Habitat: Wetlands to upland woods of the coastal plain and piedmont from Georgia to Ontario. Zones 3–7.
Propagation: Same as *A. arborea*.

At least eight serviceberries are native to the region that is the focus of this book. Even botanists have trouble telling many of them apart, both because of their similarity and because they naturally hybridize with each other. In fact, characteristics of varieties *arborea* and *laevis* get so confused that in nursery catalogs hybrids are often just referred to as a separate species, *Amelanchier grandiflora*.

If you want to impress your friends, and possibly be correct, call any bigger serviceberry tree—particularly in the southern mountains, where they can grow up to 60 feet tall—*A. arborea* var. *laevis*. Multi-trunked small trees up to a height of about 25 feet—particularly in the piedmont and upper coastal plain—are probably *A. arborea* var. *arborea*. The tall shrubs that flower when shad run in New England rivers (thus the name shadbush) are most often *A. canadensis* anywhere north of New Jersey and throughout the piedmont. In the coastal plain from Georgia to New Jersey, a stoloniferous amelanchier, *A. obovalis*, is common. *A. canadensis* and its selected forms, such as 'Prince William', are much better behaved in the landscape than aggressively colonizing *A. obovalis* and have a more pleasing aspect.

For most gardeners, just call the plant serviceberry. An eastern spring would not seem right without delicate, often fleeting white and light pink sarvice flowers catching the eye during the last gloomy days of winter. Juneberry pies were treasured by colonists and Indians alike, and were usually accompanied by stories of battles with birds. Unless you take special precautions, simply enjoy the abundance of birds as the purplish-black fruit disappears in late June. Human harvests of Juneberries are rarer than avian harvests.

Many stories exist to account for the origin of the name "service" or "sarvice." One I've heard frequently in the North relates to the fact that the earth freezes solid in the winter. Therefore, anyone who died in the winter had to await the thawing of the earth for burial. The burial "service" could finally be held when the amelanchier bloomed. Another explanation was given to me by a tiny old lady who had sat quietly through one of my lectures in the Georgia mountains. After the lecture, she slipped me a folded, handwritten note as she exited. It read, "My grandmother told me that the 'Sarvice' fruit was used to make wine for the church 'sarvice' because it is the first fruit of the year and they always ran short of wine and were anxious to get more as soon as possible. They pronounced service as 'sarvice' all the time." Take your pick of stories or create one of your own.

The final landscape reward of serviceberries is excellent and frequently enduring fall foliage color that ranges from old gold to brilliant red-orange. *A. grandiflora* 'Autumn Brilliance' was chosen for its outstanding fall color. A bonus I've noticed in trials is that rabbits are less fond of chewing on young 'Autumn Brilliance' than on other amelanchiers I've tried to grow into trees while rabbits wanted to prune them into dwarf shrubs.

I prefer placing serviceberry well away from paths where I regularly stroll. They are subject to many of the insect and disease problems that plague other members of the rose family and therefore tend to be messy at times, though the damage done to them by pests is usually cosmetic rather than life threatening. Try placing them against a backdrop of dark evergreens or water if accenting spring bloom is your landscape goal. Locations catching the afternoon sun capture the warm fall foliage colors best.

Aralia spinosa
Devil's walking stick, Hercules club

Habitat: Ravines, bluffs, slopes, upland woods, and anywhere else it can get a foothold, Florida to New Jersey. *A. spinosa* is a pioneering species in plant succession, often disappearing as the forest develops around it. Zones 4–9.

Figure 24. *Aralia spinosa*.

Propagation: Seeds, ripening in black fruit during early autumn, require 2–3 months' stratification. Remove seeds from the pulpy fruit and dry them before storing at 40° or stratifying.

Late fall root cuttings are easy, but plants sucker readily and are easily grown from divisions. Wear heavy gloves and long sleeves when dealing with this thorny beast.

There is nothing subtle about devil's walking stick. Enormous (3–4 feet in diameter) dark blue-green leaves are borne at the ends of thorny stalks 10 to 20 feet tall. Topping these vicious green umbrellas are creamy white flower clusters up to 3 feet tall. Following the mid-summer floral display, the plants are crowned with an abundance of black fruit, often with bright pink fruit stalks, until after Columbus Day.

This large, aggressively suckering shrub or small tree will grow in the poorest of soils in exposed sites. If space in your garden is at all limited, leave this one in the wild. Although it can create a dramatic, subtropical effect, I don't think it is worth the physical pain required to keep it in check.

Aronia arbutifolia
Red chokeberry

Habitat: Moist soils, springheads, and swamps from Florida to Nova Scotia. Zones 4–8.

Propagation: Seeds should be removed from red ripe fruit in mid-autumn. They require 3 months' cool stratification. Softwood cuttings root without hormone treatment. However, treatment with .5 percent IBA powder produces more abundant roots.

Figure 25. *Aronia arbutifolia*.

If you have the space for a tall, somewhat leggy shrub near wetlands and would enjoy bright red fall foliage, brilliant red berries, and the birds they attract, choose *Aronia arbutifolia*. My choice would be the selection 'Brilliantissima', which has better foliage as well as glossier and more abundant fruit than most wild plants of this species.

Space is critical. Rarely do you see just one plant. Because of red chokeberry's suckering habit, colonies develop quickly. Like many other members of the rose family, chokeberries are occasionally made unsightly by diseases and insects. For this reason, I suggest planting them where they can be seen at a distance or enjoyed in an informal or natural rather than an intensely maintained area.

Baccharis halimifolia
Groundsel tree, salt marsh elder

Habitat: Coastal plain and occasional shorelines or piedmont marshes from Florida to Massachusetts. Zones 5–9.

Propagation: Seeds are tiny. If treated like those of azaleas and kept moist, they germinate in 7 to 10 days at 75°.

Most plants in this book have showy flowers, foliage, or fruit, but *Baccharis* is included only because of the numerous questions that I got as a county agent—and that agents still continue to get—each fall, asking, "What's that plant covered with the silky floss?"

Female plants are covered with shiny white clouds beginning in early fall each year. The first plants I noticed were in Florida salt marshes; but since then, I've seen

Figure 26. *Baccharis halimifolia*.

Figure 27. *Callicarpa americana.*

plants as far north as New England, always in wetlands or at the edge of wetlands.

Although I've rarely seen groundsel tree, one of the few woody members of the daisy family, used intentionally in a landscape, it seems like a natural choice where brackish or salty water may be a problem. Perhaps it should be investigated by departments of transportation. The 10-foot shrub (I've never seen one I would call a tree, despite the common name) can create unusual interest in places where a bit of landscape diversity might be welcome.

Callicarpa americana
American beautyberry, French mulberry

Habitat: Woodlands, often in moist shade at the edge of pinewoods, Florida to Maryland. Zones 7–10.

Propagation: Seeds are easily squeezed from fruit after the first killing frost. While no scarification or stratification is required, my experience has been that fresh seeds don't germinate for about 3 months. A period of warm after-ripening is apparently needed. This can be accomplished by storing cleaned seeds at room temperature and then sowing the following spring. Softwood cuttings root easily under mist with no hormone treatment.

C. americana's most common fruit color is a traffic-stopping, iridescent pinkish-purple to light red-violet. For this reason, beautyberry seems impossible to blend into anything but an earthy color scheme of woodlands and grasses. It clashes with almost everything else. However, Edith Eddleman somehow makes it work with asters and goldenrods in the renowned perennial border at the North Carolina State University Arboretum in Raleigh.

By early September, the fruit is contrasted with yellow-green foliage to create startling visions. It is most effective at the edge of a pine woodland or at pondside, where tall grasses may call for some excitement. The white variety *lactea* grows well on the edge of deep shade, where the abundant pearly fruit is welcome in the heat of late summer. Unlike common *C. americana* fruit, that of *lactea* complements nearly everything.

The shrub itself has a loose, open character with mid-summer flowers of blush pink. Since flowers and subsequent fruit are on the current season's growth, hard pruning each winter can limit growth without limiting berries. By cutting mine nearly to the ground each year, I keep it at about 4 feet in both height and width. Otherwise, it would be nearly twice that size.

Calycanthus floridus
Carolina allspice, sweet shrub, strawberry shrub

Habitat: In openings and at the edge of woodlands, Florida to Pennsylvania. Zones 4–9.
Propagation: Kidney-bean-sized, dark mahogany brown seeds hang inside leathery, shot-pouch-appearing capsules into the winter. I have collected viable seeds in January. Texts suggest 3 months' stratification, but I've had seeds germinate with only 1 month's stratification. The triangular cotyledons on recently germinated plants must contain a powerful slug attractant. Slugs will climb over other delicacies to get to sweet shrub seedlings.

Great clonal variation exists in rootability. Try semihardwood cuttings with .8 percent IBA powder. IBA liquid in alcohol seems toxic, so the problem must be the alcohol. If you have the right plant, nearly all will root. If not, try another plant.

Figure 28. *Calycanthus floridus*.

Sweet shrub is grown for its unusual flower and haunting fragrance, which has been described as part banana and part strawberry in one text, as a combination of pineapple and grapefruit in another, and as Juicy Fruit chewing gum in yet another. In truth, the fragrance is highly variable; in some plants it is overwhelming, while other plants are scentless. Flowers are more consistent than fragrance, being mahogany purple to dark rusty brown, resembling a tiny cone more than a flower. A delightfully scented selection, 'Athens', has yellow-green flowers.

Selections with pleasant fragrance should be located near windows or outdoor living areas, where flowers, borne from mid-spring into summer, will reward people nearby. The plant itself is carefree, reaching a height of about 8 feet and spreading approximately 10 feet wide, with glossy green leaves. In the warmer parts of Zone 7 plus all of Zones 8 and 9, sweet shrub needs light shade, but further north or in higher elevations it will thrive in full sun.

Figure 29. *Catalpa bignonioides*.

Catalpa bignonioides
Southern catalpa, Indian bean

Habitat: Originally native to wetlands along the Gulf coast, this tree has been introduced from Florida to New England. Zones 5–9.
Propagation: Seeds, inside beanlike pods that often reach a foot in length, require no scarification or stratification. Both hardwood and softwood stem cuttings as well as root cuttings have reportedly been used to propagate catalpa asexually (if you can find a tree worth the effort). When asked, most landowners would probably welcome any inquiry about transplanting seedlings or root suckers away from their property. Once trees are established, they are nearly indestructible.

There are few professional garden designers who don't sneer at catalpa. It is coarse and often messy. The leaves are large and smell bad when crushed. Clusters of spotted white flowers, which are occasionally spectacular above the leaves of 25- to 40-foot trees, appear briefly in mid-June before strewing the earth with decaying petals.

Despite its drawbacks, catalpa is often found near ponds and streams throughout the South because it is the only source of a green-and-black-striped caterpillar that serves as super bluegill and shellcracker bait. These caterpillars can completely defoliate trees, but trees recover, growing a new set of leaves within a month. The caterpillars are fussy eaters. If you don't introduce them to your tree, you may never have any. Introduce a dozen and you'll have a fishbait factory beginning the next year. Perhaps this is why Virginia colonists planted catalpa in rows along Palace Green in Williamsburg.

Ceanothus americanus
New Jersey tea, mountain snowball, redroot

Habitat: Florida to Maine, usually in hot, dry areas where little else grows. Often seen at the base of south- or west-facing road cuts. Zones 4–8.
Propagation: Hot water soaks and stratifica-

Figure 30. *Ceanothus americanus*.

tion are reported to improve seed germination. I have no experience propagating this plant from seeds but suspect that fall planting or two months' moist, cool stratification would suffice. Softwood cuttings treated with .3 percent IBA powder root readily in summer.

New Jersey tea is a neat, low-growing (up to 3 feet) shrub that is awaiting discovery. Perhaps this will be the decade when the inevitable superior clone will be discovered and popularized.

Covered in late spring with puffy plumes of white to light pink scentless flowers (this plant is sometimes called mountain sweet, but I've never noticed any fragrance), *Ceanothus americanus* is a problem-solving shrub. It tolerates hot, dry conditions and is therefore often found in excessively drained soils, on sun-baked south slopes, and at the edges of dry woodlands. The large, deep red, woody roots as well as the leaves were used for tea in colonial times. A lotion made from the leaves is supposed to remove freckles.

Mass plantings, in full sun or light shade, can provide quite a show in areas where it is a challenge just to get other plants to survive. Unfortunately, this *Ceanothus* is the only one that seems to tolerate our eastern humidity. Other species and hybrids provide some of the finest blue flowers in the western United States and England but usually don't survive eastern summers, even where they are hardy. Research is currently underway to investigate using New Jersey tea as a rootstock for these superior blue beauties in eastern gardens.

Cephalanthus occidentalis
Buttonbush

Habitat: Along the edges of wetlands, frequently growing in shallow water, from Florida to Canada. Zones 5–10.
Propagation: Seeds, actually nutlets, are carried in clusters well into the winter. They require no special treatment. Softwood or hardwood cuttings root without hormone treatment.

Figure 31. *Cephalanthus occidentalis*.

Buttonbush is intolerant of dry soils. However, if you have a site that is uniformly wet and you wish to feed waterfowl, consider buttonbush. The white buttonlike flowers appear in mid-summer on younger, more vigorous stems. Therefore, to encourage flowering and keep this unkempt shrub from achieving its maximum height of about 10 feet, prune it all the way to the ground every 2 or 3 years. Patiently await new growth, as buttonbush is one of the last plants to flush leaves in the spring. Pruning won't kill the plants, but slow regrowth may scare gardeners.

Cercis canadensis
Redbud, Judas tree

Habitat: Grown from Florida to New England but in the wild seen most often along fencerows on moist and slightly acid soils in the lower mountains, piedmont, and coastal plain. Zones 4–9.

Propagation: Seeds require scarification followed by 60–90 days' stratification. The simplest way to grow redbuds is to scarify seeds and fall plant outdoors. Cuttings are nearly impossible. I've never rooted one.

Figure 32. *Cercis canadensis.*

Redbud is a welcome bit of nonwhite color among early blooming native trees and shrubs. For this reason alone, it will continue to be popular. The pinkish-purple, pea-shaped flowers provide a promise of things to come on the many gray, rainy days of early spring.

Redbud is a legume, meaning it can manufacture its own nitrogen fertilizer when grown in the right soil. For this reason, acid soils should be limed rather than fertilized in the immediate vicinity of a redbud tree. The heart-shaped leaves characteristic of this small to medium-sized deciduous tree seem to disappear into the landscape after flowering and produce little color of merit in the fall.

Redbuds are best placed where they will be seen in the spring, with flowers contrasted against woodlands or conifers. They do best in moderately to richly fertile, well-limed soils but will survive in far poorer sites. Selections have been made with white, pink, and double flowers, and with purple or with white and green variegated leaves. These unusual types are gradually making their way to the marketplace thanks to the remarkable grafting skills of a few dedicated horticulturists.

Native redbuds tend to be short-lived, reaching their peak attractiveness shortly after arriving in the landscape and continuing for about 20 years. Thereafter decline can be abrupt, often due to canker in moister sites. The only appropriate cure for canker is to remove the tree. The other major problem with redbuds is heavy seeding, which can lead to redbud as a weed.

Chionanthus virginicus

Fringe tree, old man's beard, Grancy Greybeard

Figure 33. *Chionanthus virginicus*.

Habitat: Grown all the way to Canada but native from Florida to New Jersey. Found on dry, acid soils near moisture, such as on streambanks or slopes above wetlands. Zones 3–9.

Propagation: Seeds are inside awful-smelling blue fruit, which ripens in early fall. Remove seeds from fruit before planting. Seeds have a double dormancy that requires a full season (90–150 days) of warm, moist treatment for roots to emerge from the seed and then another 30–60 days of cool stratification for the tops to sprout. Fall-sown seeds germinate the second spring after planting. Better propagators than I have been unable to root fringe tree cuttings.

Fringe tree is a large shrub or small tree, up to about 25 feet tall, worthy of being placed as an isolated specimen or mixed into a border. Its foliage is deciduous, medium to dark green on the upper surface and lighter below. The primary attraction is the delicately fragrant flowers, which appear on last year's growth in pendulous ivory to pure white clusters 6–8 inches long just as the leaves unfold in late spring, about Mother's Day in the Asheville, North Carolina, area or mid- to late April in Atlanta. Male trees generally have more dramatic flowers than female.

Fringe tree has a reputation for being difficult to transplant, but I've seen no problem with plants that are moved while fully dormant and then cared for rather than abandoned the first year in the landscape. Perhaps the problem has been in choosing a site, since fringe tree clearly prefers fertile, slightly acid soils with average moisture. Although native sites may appear dry, they are not droughty. Choose your site accordingly, and be patient. Seedlings won't flower for 5 to 7 years under the best of conditions. This slow grower is well worth the wait.

Fringe tree has for some reason been slow to become appreciated. John Bartram introduced it to England via Peter Collinson in 1736. Some 195 years later it received an award of merit from the Royal Horticultural Society.

Cladrastis lutea (kentuckea)

Yellowwood, gopherwood, vergilia

Habitat: Rich cove soils of upland North and South Carolina and Georgia. Zones 5–8. Uncommon.

Propagation: Like the seeds of many leguminous trees, those of yellowwood need scarification. Ninety days' stratification is reportedly needed, but don't believe it. If seed coats are thoroughly scarified, you can skip stratification.

Root cuttings taken in December and held in nearly dry sand have proven successful. Transplanting to well-drained, fertile soil or soilless media in mid-spring, followed by low rates of fertilizer the first year, were required to continue the success. Root cuttings seem more fertilizer-sensitive the first year than seedlings.

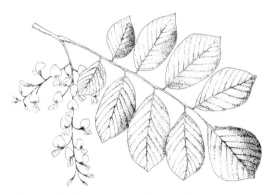

Figure 34. *Cladrastis lutea (kentuckea)*.

Yellowwood is named for the color of its heartwood. However, the name doesn't do this fantastic flowering ornamental tree justice. Once the problems of nursery production are solved, this will become one of the most widely sought landscape trees.

Among its attributes are that it provides excellent shade but has such deep roots that grass and shrubs grow easily beneath it. Picture a tree with dark green leaves, 50 feet high and just as wide but with growth slow enough that it doesn't become brittle and break up. Now cover it with pendulous foot-long, wisterialike clusters of white fragrant flowers in late spring. In the fall, foliage turns delicate orange or yellow before dropping. Once established, it is drought tolerant and nearly pest free.

Why isn't yellowwood more popular? Slow growth in early years is part of it. Difficulty in transplanting larger specimens due to the deep roots adds to the problem. Too often soils have not been adequately limed, so landscape trees struggled to become established. Flowers may not appear until the trees are 10 or more years old, then only appear in quantity every 2 or 3 years. If pruned in the spring, yellowwood will "bleed" excessively, and if pruned immediately after transplanting, it may not survive. Despite all this, plant a small tree and be patient. Yellowwood is one of the gems among native flowering trees.

Clethra acuminata
Cinnamonbark clethra, mountain pepperbush

Habitat: Shady streambanks and springheads as well as dry-appearing hillsides in the mountains and upper piedmont from Georgia to Pennsylvania. Zones 5–8.
Propagation: Despite appearances, seeds don't mature in the dried flower heads until long after leaves have dropped, not until early November in the North Carolina mountains. Handle like azaleas, as seeds are tiny and require no pretreatment to germinate.

Softwood cuttings root in well-drained media under mist when treated with a 1,000 ppm IBA solution. I've rooted cuttings in good percentages without hormone, but plants grown from these cuttings seem to lack vigor. Great variation in rootability

Figure 35. *Clethra acuminata*.

exists from plant to plant. If you fail with one plant, try another.

Cinnamonbark clethra is about to be discovered by gardeners everywhere. When grown in the deep, moist shade of its native habitat, plants are often leggy and flowers unnoticed. However, when moved to a drier, sunny site, cinnamonbark clethra becomes a wonderfully valuable shrub or small tree. White, lily-of-the-valley-like flowers rise above and hang from terminals contrasting with medium green foliage during early summer when little else is blooming. Since flowers open from the base to the tips of 6- to 8-inch racemes, plants stay in attractive bloom for weeks. Fall color is a fleeting old gold. Once temperatures get down to 25° or lower, foliage browns and drops rapidly.

If you allow your cinnamonbark clethra to reach its full potential, pruning away lower branches and suckers, the reason for its name becomes apparent. A trunk color from pinkish-gold to deep cinnamon red is revealed as bark naturally peels away. Variation from one plant to another is enormous. Small tree-form plants 15 to 20 feet tall could be the perfect accent for your patio. If mid-summer flowers are your goal, keep plants pruned to well below eye level, as each year's new growth will add a couple of feet to the shrub's height before flowering.

Clethra alnifolia
Summersweet, sweet pepperbush

Habitat: Coastal plain and piedmont wetlands from Florida to New England. Zones 3–9.
Propagation: Same as *C. acuminata*.

Figure 36. *Clethra alnifolia* 'Rosea'.

Summersweet is a versatile, desirable shrub which responds well to pruning. Its ultimate height can be 10 feet. When planted in moist soils or partially shaded locations, it is almost carefree, with none of the messy character of some other summer-flowering shrubs. Allowed to grow, it is a dependable background or hedge plant, one of the few that will tolerate moist soils and provide fragrant flowers.

Culturally, clethras can be treated as azaleas, but they are actually more tolerant of higher soil pH and fertilizer. If you can find them, *Clethra alnifolia* selections are worth considering. 'Paniculata' has larger flowers than other selections; 'Rosea' has pink flowers that fade to white, with excellent habit and dark green, glossy leaves; and 'Pinkspire' is a slightly smaller form having rose-pink buds that open to a pink which does not fade. Perhaps the best white summersweet for the garden is the dwarf selection 'Hummingbird'. It was discovered by Fred Galle at Calloway Gardens in Georgia but has been popularized by a nursery near Philadelphia, which reports good hardiness. All of the summersweets should be sought by coastal gardeners since they have excellent salt spray tolerance.

Cornus alternifolia
Pagoda dogwood

Habitat: Well-drained clearings or woodland edges from the Georgia mountains to Canada. Zones 3–7.
Propagation: Seeds mature inside blue-black fruits during late summer. Cleaned seeds require 5 months' warm stratification followed by 3 months' cool stratification to germinate.

Softwood cuttings can be rooted using .8 percent IBA powder but must not be disturbed until new vegetative growth has been forced by supplemental lighting or cuttings have been allowed to go through a normal winter dormancy and have begun vegetative growth the following spring.

Figure 37. *Cornus alternifolia*.

Pagoda dogwood is the only alternate-leaved dogwood native to North America. Greenish-white, unpleasant-smelling (at least to me) flowers are borne in layers in late spring. By late July, the fruits are turning from green to a dark blue-black on red stems.

There are three reasons for considering this native: (1) It usually achieves a height of only 15–25 feet but is often half again as wide as it is tall, providing needed horizontal lines in highly vertical landscapes. (2) It will survive cold that would kill flowering dogwood, *C. florida* (anywhere the temperature regularly falls below −20°, flowering dogwood probably isn't worth the trouble). (3) It has excellent resistance to dogwood anthracnose disease.

Cornus amomum
Silky dogwood, silky-cornel

Habitat: Edge of wetlands and on streambanks from Florida to Canada. Zones 5–8.
Propagation: Harvest fruit when it turns black. Remove seeds from fruit and stratify 90 days. Cuttings of silky dogwood are the

Figure 38. *Cornus amomum*.

Cornus florida
Flowering dogwood

Habitat: Understory tree in open woods and at the forest edge, Florida to New England. Listed as Zones 5–9 but at its best in Zones 6–8.

Propagation: Seeds should be removed from red fruit in fall and then stratified for 90 days. Most nurserymen sow cleaned seeds outdoors from late October through Thanksgiving, letting seeds stratify naturally. Cuttings should be handled like those of *C. alternifolia*.

easiest to root of the dogwoods listed. IBA powder (.1–.3 percent) works well on semi-hardwood cuttings; no hormone seems necessary on softwood cuttings.

Silky dogwood will tolerate poorer drainage than flowering or pagoda dogwoods but can't compare with them as a valuable landscape specimen. Whitish flower heads are relatively flat on top, appearing in mid-spring beside the earliest elderberries in moist roadside ditches and at the edge of marshes and wet woodlands. Fruit is pale blue, turning black as the season progresses.

No amount of pruning will convince this dogwood to be a tree. Its form is that of a coarse shrub, even though specimens 10–15 feet tall are not uncommon. In wet areas or those too cold for other white-flowering wetland shrubs, silky dogwood might be something to try.

Flowering dogwood is the most popular landscape tree sold in the eastern United States, with good reason. Flowering dogwoods turn suburban landscapes and rural woodlands into a fairyland of pink and white blooms early each spring. The red berries attract acrobatic squirrels and flocks of birds each fall. In between, flowering dogwood is a well-behaved, 25- to 30-foot-tall member of the community.

Failures with flowering dogwoods almost always come from planting in the wrong place. They thrive in soil that is at least moderately drained, slightly acid (add 50 pounds of dolomitic limestone per 1,000 square feet every 3 years in most areas), and partly shady. *C. florida* is an understory species that will *tolerate* heavy shade in the Southeast but not thrive and bloom well unless it receives a few hours of direct sunshine or the dappled sunlight of high pine shade each day. To protect against dogwood anthracnose disease in cooler parts of the region, at least a half day of sunshine and good air drainage are essential for flowering

Figure 39. *Cornus florida*.

dogwoods. Because of flowering dogwood's shallow root system, grass should be kept completely away from the base of the tree at least until the plant is well established.

Fortunately, growers are developing selections that are more resistant to diseases other than dogwood anthracnose and are better adapted to local climates. 'Cherokee Princess' is a favorite white selection because of its excellent vigor and disease resistance. 'Cherokee Chief' and 'Rubra' are by far the most popular red and pink varieties sold. Pink and red varieties should be purchased on the advice of the local garden center if you live along the Gulf, where some varieties don't perform as well as others. Progress is coming gradually in producing dogwoods with colored blooms that flower consistently in the warmer parts of Zone 8 and in Zone 9.

In the twentieth century, the economic importance of dogwood outside the landscape has almost been forgotten. The tree's name comes from the old word "dag," meaning skewer. As the name suggests, this hard, tough, splinter-free wood was used in making skewers to hold meat together while cooking. It was also used in numerous other wood products, including weaving shuttles, spindles, mallet heads, chisel handles, and pulleys.

Crataegus phaenopyrum
Washington hawthorn

Figure 40. *Crataegus phaenopyrum*.

Habitat: Moist areas from Florida to Pennsylvania, often forming thickets. Zones 5–8.
Propagation: Washington hawthorn seeds are supposed to have a double dormancy, requiring 4 months' warm stratification followed by 4 months of cool stratification. However, the few I've tried have germinated following cool stratification, without benefit of warm stratification, so don't be surprised if seeds germinate the spring following planting. Remove seeds from fruit before treatment but don't store them for more than a year as they lose viability rapidly.

A dozen species of hawthorns are native to the eastern United States. All have flowers that resemble those of apples or cherries, depending upon your perspective, but the fragrance is far less appealing than that of apple blossoms. The telltale characteristics identifying this plant for the uninitiated are spikelike thorns, abundant fruit ("haw apples"), and the presence of bird nests.

I've chosen Washington hawthorn to represent the genus because it has been a consistent performer in landscapes from Asheville, North Carolina, to New England. The mid-spring flowers are brilliant white. The fruit is bright red and persists well into winter, depending upon the local avian population. Birds love this small tree, which gives them cover, a spot to build nests, and a source of food.

The most effective landscape uses for hawthorns vary with individual needs. They can form an effective, almost impenetrable, windbreaking screen. The strong horizontal branching and gray bark provide a landscape substance similar to that of beech. Therefore, this 25-foot tree can be used to contrast with a building or be viewed by bird lovers from a window. The excellent, slightly larger selection by Robert Simpson of Vincennes, Indiana, 'Winter King', may be a hybrid between Washington hawthorn and another native, the green hawthorn, *C. viridis*. It is usually listed as *C. viridis*, however.

Cyrilla racemiflora
Leatherwood, titi

Habitat: Along streams as well as in bay woods and swamps, Florida to Virginia. Zones 6–10.
Propagation: Seeds will often hang on plants well into winter. They require no special treatment. Semihardwood cuttings treated with .8 percent IBA powder root in high numbers.

Figure 41. *Cyrilla racemiflora.*

I first encountered titi as a potential summer nectar source when working with beekeepers in Florida. But it can be a valuable shrub in natural waterside landscapes whether you keep bees or not.

Its rich green foliage can provide a perfect backdrop to herbaceous summer flowers or serve to accent its own abundant drooping white flower clusters that bloom anytime from late spring into summer. In Zones 6 and 7 and the cooler parts of Zone 8, the thick, lustrous leaves will turn orange or red in the fall. Further south, plants are nearly evergreen.

Titi is not particularly aggressive, so it doesn't tolerate competition from other shrubs and trees very well. In other words, it will stay where it is put as long as you can ensure moist roots. Titi doesn't do well in dry soils.

Diervilla sessilifolia
Bush honeysuckle, mountain honeysuckle

Habitat: Mountains and piedmont, usually in dry, sunny, exposed sites, Georgia to West Virginia. Zones 4–8.

Propagation: I'm sure this plant can be grown from seed, but I've never tried to do so. Mike Dirr writes that they have no special requirements for germination. Cuttings, both softwood and semihardwood, root readily.

Bush honeysuckle is a 3- to 5-foot shrub with dark green leaves, often with red veins and stems. Yellow flowers are borne on the tip of new growth and, while small, are abundant all summer. I've never noticed any fragrance, but they still manage to attract pollinating insects.

If you have a sunny site where nothing else will grow, try bush honeysuckle. It thrives in windy, sunny, dry locations of the mountain South and in the hills further north. I've never seen an insect or disease become a real problem on bush honeysuckle.

This plant is best when pruned back in late winter or early spring to remove old and winter-killed growth that may be unsightly. It is not tolerant of shady conditions but will sucker and establish colonies in full sun.

Figure 42. *Diervilla sessilifolia.*

Diospyros virginiana
Persimmon

Habitat: Abandoned agricultural lands, fencerows, open woodlands, Florida to southern New England. Zones 4–9.

Propagation: Seeds should be extracted from ripe fruit in mid-autumn. They require 90 days' stratification. If you find a truly superior tree, grow some seedlings for a year or two; then find someone who knows how to chip bud scions from the superior tree onto your seedling rootstock. Late winter chip budding and veneer grafting have worked when trying to propagate persimmons asexually. Cuttings have not.

Figure 43. *Diospyros virginiana*.

The value of our native persimmon may be greater for wildlife and history than for modern landscapes. Most of the plants I've seen were in old fields or sites where little else would grow. Persimmon is a tough plant that is a large shrub just as often as it is a tree.

Colonial literature is full of references to persimmon pudding or jelly, with some suggestions that ripe fruit added to the fermentation pot will enhance home-brewed beer.

Courtesy of June T. Smith of Tuckasegee, North Carolina, here's the recipe for the best persimmon pudding (with a consistency closer to that of a moist, cakelike date bar) I've ever tasted.

- 2 cups persimmon pulp
- 3 eggs, beaten
- 1¾ cups milk
- 2½ cups flour
- ½ tsp. baking powder
- ½ tsp. baking soda
- ½ tsp. salt
- ¾ tsp. cinnamon
- ¾ tsp. nutmeg
- 1⅔ cups sugar
- ¾ stick butter, melted

Collect about a quart of persimmons, wash carefully, and extract the pulp by pressing them through a colander, food mill, or Squeezo strainer. If the pulp is very thick, add enough water to make it the consistency of pancake batter. You should have about 2 cups of pulp. Add the eggs and milk to the persimmon pulp and mix. In a separate bowl, sift together the flour, baking powder, baking soda, salt, cinnamon, nutmeg, and sugar. Combine the dry ingredients with the pulp, eggs, and milk. Add the butter and stir all together. Pour into a well-greased 9 × 13 baking dish. Bake at 325° for one hour.

Reserved parishioners fight over this persimmon pudding at October church suppers.

The hard, dark wood of persimmon—a first cousin is ebony—was used for utensils and later for the heads of golf clubs. Deer, foxes, and opossums treat the ripe fruit as a delicacy, perhaps because they know how to harvest it. Only dead ripe fruit should be eaten. Connoisseurs suggest spreading a sheet under trees on the autumnal equinox and then checking for ripe "drops" daily. Do not shake the tree, and discard any "drops" that are very firm. Collect and freeze the daily harvest until you have enough fruit for multiple batches of your favorite persimmon recipe.

If you want a native persimmon tree, planting any place in the full sun will do. Be sure to transplant while the tree is young, as the deep tap root makes moving larger plants chancy. Persimmons are usually either male or female, with an occasional tree both. When they are young, unfortunately, you can't tell which you've got.

Euonymus americanus
Hearts a bustin', strawberry bush

Habitat: Moist woodlands, streamsides, and colonial landscapes from Florida to New York. Zones 6–9.
Propagation: Seeds given 90 days' warm stratification followed by 60 days' cool stratification will germinate, but my percentages have always been low. Semihardwood cuttings root readily with no hormone treatment.

Much of the year, this open, 6-foot-tall shrub, attracts little attention except on the part of deer and rabbits who enjoy euonymus salad occasionally. However, when the warty red fruits open in late summer to reveal shiny orange to scarlet seeds it is nearly impossible to ignore.

Although hearts a bustin' will tolerate

Figure 44. *Euonymus americanus*.

very deep shade, it fruits best when grown in light shade with a minimum of fertilizer. Placed where it must be passed on early autumn strolls, it will make walkers slow or halt and be sure to comment. Usually the comments run to "What's that?" or "My grandmother used to have one of those. What's its name? Something to do with love."

Fothergilla gardenii
Dwarf fothergilla

Habitat: Sunny, higher ground in swamps and adjacent moist grassy areas of the coastal plain from Georgia to North Carolina. Zones 5–8.
Propagation: Seeds require at least a 6-month warm, moist period followed by 3 months' cold stratification and can be tricky. I don't understand why fothergilla is supposed to be difficult to root. I've never had any problems. Semihardwood cuttings root relatively easily with or without hormone. Rooted cuttings should not be moved until

Figure 45. *Fothergilla gardenii* 'Blue Mist'.

new vegetative growth is over an inch long or normal winter dormancy is completed.

See the comments following *F. major* below.

Fothergilla major
Large fothergilla

Habitat: Sunny, dry uplands of the southern Blue Ridge Mountains in Georgia, North Carolina, and Tennessee. Zones 4–8.
Propagation: Same as *F. gardenii*.

Both of these fothergilla species are listed as endangered, yet they fit beautifully into a woodland garden of azaleas and rhododendrons. Both are relatively easy to grow if treated like azaleas. However, avoid very dry sites at all costs. Fothergillas need moisture to survive.

F. gardenii is supposed to be 3 feet high, while *F. major* grows up to 10 feet tall. I've seen *F. gardenii* grow taller when fertilized and grown in light shade. If I wanted a 4-foot fothergilla, my choice would be *F. major* for larger flowers, greater hardiness, and drought tolerance. I'd just use my pruners to keep it at the height I want. The hard part is deciding when to prune. Prune in summer if you want to emphasize flowers. Prune in late winter if you want to emphasize foliage.

Both species have distinctive, brushlike, creamy white flowers. Fragrance is highly variable. Both can possess one of those pleasant, lingering fragrances that stir memories of previous woodland strolls. The best have a honeylike scent lighter than tulip poplar and heavier than sourwood honey. The fragrance of sweet clover honey is close to, but not exactly the same as, that of the most pleasantly fragrant fothergilla flowers. Some fothergilla flowers are scentless or musky.

Perhaps the finest feature of these unusual flowering shrubs is their brilliant fall foliage. For this reason, plant them where the afternoon autumn sun will illuminate the orange, red, and yellow on each leaf. To get the best floral and foliage display, give fothergillas as much sunshine as you can manage. The blue-leafed *F. gardenii* selection 'Blue Mist' or the vigorous, floriferous selection 'Mount Airy', a natural hybrid of these two species, may finally bring this worthwhile native to prominence.

Gordonia lasianthus
Loblolly bay

Habitat: Wetlands, Florida to North Carolina. Zones 8–9.
Propagation: Seeds require no special treat-

Figure 46. *Gordonia lasianthus*.

ment. Softwood and semihardwood cuttings root readily with low rates of hormone or none.

The dark green glossy foliage of *G. lasianthus* provides the perfect backdrop for numerous 2- to 3-inch-diameter waxy white flowers throughout the summer and then stays evergreen. Gordonia is useful in coastal and warmer piedmont landscapes as either a large shrub or small tree.

Loblolly bay has a reputation for being difficult to grow, but it has been used successfully at Walt Disney World in Florida and in restored areas along the southeastern coast. If you remember that it is allergic to drought and won't tolerate a lot of fertilizer at any one time, you should have little problem with this terrific showy native.

Halesia carolina (tetraptera)
Carolina silverbell

Habitat: Along waterways or in moist, rich woodlands from Florida to Virginia. Zones 5–8.

Propagation: Seeds have double dormancy requiring 4 months of warm stratification followed by 2 to 3 months of cold stratification. Sometimes a few seeds will germinate

Figure 47. *Halesia carolina*.

with only cool stratification. Even when everything is done right, I've never gotten much more than 25 percent germination.

Softwood cuttings treated with .8 percent IBA powder or quick dipped in 2,500 ppm IBA liquid root in high percentages. Rooted cuttings should not be transplanted until growth flushes the spring following rooting or a flush of vegetative growth is forced with artificial lighting.

See the comments following *H. diptera* below.

Halesia diptera
American snowdrop tree

Habitat: Rich woods at the edge of wetlands from Florida to South Carolina. Zones 5–8.
Propagation: Seeds require 3 months' cold, 6 months' warm, and then another 4 months' cold stratification. As with *H. carolina*, an occasional seed ignores the rules and germinates with only cool stratification. Treat cuttings as recommended for *H. carolina*.

Both *H. carolina* and *H. diptera* can create a fairyland effect in streamside woods. The delicate white or pale pink bell-like flowers hanging from twiggy branches are dramatic only when plants are used as specimens where the striped bark of younger trees enhances interest. Their effect is usually far more subtle. Carolina silverbell blooms a week or so before dogwoods, while most years the shrubby snowdrop blooms at the same time as dogwoods, so their subtlety is often missed due to the overwhelming bracts of flowering dogwoods.

Carolina silverbell is an excellent tree for dooryards, where a touch of spring bloom as well as delicacy is appreciated. Most silverbells never exceed 30 feet in height and have a moderate growth rate. As long as they are watered during severe droughts, silverbells seem able to tolerate other cultural indignities.

The snowdrop tree, while most often seen as a shrub, makes a more dramatic floral show than Carolina silverbell, particularly if you can find the variety 'Magniflora'. Snowdrop trees can be pruned to be delightful patio trees, 15–20 feet tall. A flowering American snowdrop tree overhanging a path

Figure 48. *Halesia diptera*.

near the greenhouses at Asheville's Biltmore House and Gardens is a traffic stopper.

Hamamelis virginiana
Witch hazel

Habitat: Rich woodlands from Florida to Nova Scotia. Zones 3–8.
Propagation: Seeds should be collected from capsules during late autumn, nearly a year after flowering. They must be collected before capsules fully mature, as ripe seeds are expelled for great distances when capsules open. Put closed capsules in a paper bag and, as they dry, you'll hear what sounds like corn popping as seed capsules open. Seeds have double dormancy requiring 2 months of warm stratification followed by three months of cool stratification.

Good results are relatively easy to obtain with softwood cuttings from other *Hamamelis* species and hybrids as long as vegetative growth is forced after cuttings have rooted by applying light fertilization and long days, but I've had no success in my few attempts at rooting the native witch hazel. I suspect that my cuttings were taken too late in the season. Reports suggest that the earlier witch hazel cuttings are taken the better.

There are two good reasons—beyond the novelty of shooting seeds—for growing the native witch hazel: the foliage on this large shrub turns brilliant yellow-gold in the fall, and unusual yellow, pleasantly fragrant flowers appear a little later in the fall, as foliage is dropping, and remain for weeks on

Figure 49. *Hamamelis virginiana*.

leafless stems at a time when little else is in bloom.

The yellow of witch hazel leaves and flowers comes alive in autumn sunlight. For this reason, it should be planted in a shrub border or at woodland's edge where afternoon sun can bring it to full splendor. Fall foliage makes a bold color splash while the curious, spidery flowers remind us that the natural world is not asleep when leaves have fallen. The fragrance is most welcome in cool autumn air.

Because of its size, up to 25 feet tall, the native witch hazel should be given plenty of space. A number of relatives and hybrids are smaller but bloom in mid- to late winter rather than fall, so their flowers are often missed.

I knew witch hazel as medicine long before I knew it as a flowering plant. In my youth it was possible to see bundled witch hazel brush cut and waiting to be taken to the distiller who would extract the medicine. Before the Dickinson family's commercial preparation was available in stores, however, an Oneida Indian supposedly discovered the plant's medicinal properties. I

Figure 50. *Hydrangea arborescens.*

wonder if the Oneidas use forked twigs of witch hazel for dowsing (water witching), another use for this native plant that has survived to today.

Hydrangea arborescens
Wild hydrangea, smooth hydrangea

Habitat: Shady mountain and piedmont roadsides from Georgia to New York. Zones 3–9.
Propagation: Seeds require no special treatment but need careful handling just because they are so small. Softwood cuttings should be taken early and treated with .1 percent IBA powder. Dormant hardwood cuttings should be treated with .3 percent IBA powder.

See the comments following *H. quercifolia* below.

Hydrangea quercifolia
Oak-leaf hydrangea, sevenbark

Habitat: Hillsides and along streams in wooded areas from Florida to the mountains of Georgia. Zones 5–9.
Propagation: Seeds should be treated like those of *H. arborescens*. Semihardwood cuttings should be taken early and treated with .3 to .5 percent IBA powder. Cuttings must have very well drained media and should not be disturbed until growth starts in the following spring. Because of wide variability in seedlings and the recent appearance of many superior selections, propagation of oak-leaf hydrangea by means of cuttings dominates the nursery industry.

Both hydrangeas usually have showy white sterile or staminate florets and do well in shaded locations. Here the similarity ends. The wild hydrangea usually has delicate, graceful flowers. Even at its mature height of 4–6 feet, it all but disappears into the landscape when not in bloom. Oak-leaf hydrangea is a powerful, showy part of the landscape that can stop traffic when bloom-

Figure 51. *Hydrangea quercifolia.*

ing in masses beneath the shade of tulip poplars or oaks. Allowed to grow, it will attain a height of 12 feet and be at least that wide as well.

The best wild hydrangea selection, 'Annabelle', has more plentiful bracts and larger flower heads than plants commonly seen in the wild. The oak-leaf selection 'Snow Queen' remains in bloom for weeks, with bright white 6- to 12-inch flower heads that can withstand heavy rains without dragging in the mud. 'Harmony' and 'Snowflake' have the largest flowers among oak-leaf hydrangeas, but the plants I've seen cannot withstand the torrential rains of our mountain summers without the immense flower heads falling over into the mud. The flower heads of oak-leaf hydrangea gradually turn rosy shades tinged with green and then brown, not falling from the plant until midwinter. This makes them excellent candidates for dried flower arrangements.

Both *H. arborescens* and *H. quercifolia* do best in light shade in well-limed soil. However, both will thrive in slightly acid soils. Wild hydrangea should be planted where a delicate accent is needed, such as near a rock outcropping or rail fence, while the boldness of oak-leaf hydrangea is sometimes difficult to blend into a landscape design. Use of the latter in shady border edges, as single specimens, or for mass plantings is appropriate. I've seen it used effectively all three ways. The fall foliage color of wild hydrangea is the yellow-brown common among many deciduous shrubs; the bold, large-lobed foliage of oak-leaf hydrangea turns to deep, rich reds reminiscent of wine.

Both have occasional insects nibbling on foliage, but neither has any real pest problems. If you wish to keep them as smaller shrubs, canes should be cut to the ground during winter every few years. Wild hydrangea is often treated as an herbaceous perennial and cut to the ground every winter. If you let these hydrangeas grow tall, the naturally peeling bark will reward you. Exfoliating bark is one of the features that drew William Bartram's attention when he discovered the oak-leaf hydrangea in the 1770s.

Hypericum prolificum
Saint-John's-wort

Habitat: Well-drained piedmont and mountain woodlands, Georgia to New York. Zones 4–8.

Propagation: Seeds require no special treatment but are rarely used since softwood cuttings root easily when treated with .1 percent IBA powder.

Thirty species of *Hypericum* are listed in Radford, Ahles, and Bell's standard reference, *Manual of the Vascular Flora of the Carolinas*. I've chosen Saint-John's-wort for its dependability in both full sun and light shade in dry as well as heavy soil types.

Saint-John's-wort rapidly forms a low, dense bush covered with blue-green, glossy leaves and bright yellow flowers from late spring into the heart of summer. Few plants will fill trouble spots rapidly but not grow out of bounds. However, this *Hypericum* will and rarely grows taller or wider than 4 or 5 feet. Since blooms are on the current season's growth, if you need a shorter plant, prune in late winter.

Saint-John's-wort is a readily available

Figure 52. *Hypericum prolificum*.

plant that is vastly underused in full-sun sites. The only real tricks to growing it are to lime the soil every few years and pull weeds in the spring. By mid-summer, weeds will have trouble competing.

Ilex decidua
Possum haw

Habitat: Upland piedmont and coastal plain woods from Florida to Maryland. Zones 5–8.
Propagation: Seeds should be removed from ripe, red fruit and then sown in a protected place. Some folks report success with stratification, but the essential ingredient seems to be time (sometimes years). Once internal conditions in the seed are right, it will germinate in warm, moist soil. Cuttings are much surer; .5 percent IBA powder on softwood cuttings or .8 percent powder on semihardwood cuttings both give excellent results.

See the comments following *I. vomitoria* below.

Figure 53. *Ilex decidua* 'Pocahontas'.

Ilex opaca
American holly

Habitat: Mostly piedmont and coastal plain woodlands, hedgerows, and old fields from Florida to southern New England. Zones 6–9.
Propagation: Seeds should be treated the same as those of *I. decidua*. Winter cuttings should be wounded and then treated with .8 percent IBA powder. Rooting percentage varies greatly from plant to plant.

See the comments following *I. vomitoria* below.

Figure 54. *Ilex opaca*.

Ilex verticillata
Winterberry, black alder

Habitat: Wetlands from Florida to Nova Scotia. Zones 3–8.
Propagation: Treat seeds as indicated for *I. decidua*. Semihardwood cuttings root readily when treated with .8 percent IBA powder or a 2,500 ppm IBA liquid quick dip.

See the comments following *I. vomitoria* below.

Ilex vomitoria
Yaupon holly

Habitat: Moist woodlands, pocosins, and sandhills from Florida to Virginia. Zones 7–9.
Propagation: Treat seeds as recommended for *I. decidua*. Semihardwood cuttings root with or without hormone, depending upon the plant. Great differences in rootability are found from plant to plant.

Limiting myself to only four native holly species for the landscape was difficult. Native hollies encompass beautiful evergreen trees, both pendulous and upright, as well as low evergreen foundation shrubs, often seen pruned into green landscape meatballs around fast food establishments. They also include both trees and shrubs that drop their leaves in the winter but retain attractive, colorful berries. The lore of hollies involves Druids, winter festivals, folk medicine, and wildlife gardening. I think most hollies have a place in the landscape. They

Figure 55. *Ilex verticillata.*

are wonderful plants, if not without problems.

The first problem for anyone seeking berries is that you need both a male and female plant to have berries. In commercial landscapes where no wild pollinator is likely, plant five females to one male. But keep in mind that just having a male holly, if it is not the same species as the females or does not overlap with them in bloom, is not enough. Fortunately, hollies are such popular landscape plants that it is worth the risk of planting a female and waiting a few years to see if there is an obliging male lurking nearby. Honeybees can fly with pollen from neighbors' yards and seem to love hollies. In fact, the famous gallberry honey results from the attractiveness of hollies (often a mix of *I. coriacea* and *I. glabra*) to bees.

I've chosen two evergreen and two deciduous hollies on purpose. The most widely recognized evergreen holly is American or Christmas holly. It becomes a gorgeous specimen tree, up to 60 feet tall and 40 feet wide in Zones 7 and 8. Treat American holly culturally as a rhododendron, giving it slightly more fertilizer after the first five years in your landscape. The light gray bark, pyramidal shape, and dark green leaves are stunning whether your tree produces berries or not. Over 1,000 named cultivars exist, so check with local nurseries to see which is best for your area.

Yaupon, my second evergreen holly, is far

Figure 56. *Ilex vomitoria* 'Pendula'.

more tolerant of soggy growing conditions and sweet (slightly alkaline) soils than is American holly. While yaupon can be a small tree, it is most often grown as a shrub. Its leaves are small, usually less than 1 1/2 inches long. They are, incidentally, high in caffeine and were brewed by the Indians to make tea. The red fruit is only about 1/4 inch in diameter but often persists until spring.

Many selections of yaupon holly exist. 'Pendula' ('Grey's Weeping') makes a striking patio or courtyard specimen. 'Schelling's' or 'Stroke's Dwarf' is an excellent compact shrub with rich dark green leaves, but, since it is a male, it produces no berries.

The deciduous hollies I've chosen can also be thought of as either tree or shrub. Possum haw makes a striking specimen when trained as a small tree. A perfect plant frames one of my favorite benches at the North Carolina Botanical Garden. The light gray, twiggy horizontal branches remain covered with berries in autumn after leaf fall. Currently, my favorite *I. decidua* cultivar is 'Warren's Red' because of its heavy fruit set and dark green foliage. 'Pocahontas' is more subtle but an equally worthwhile selection. Planted with evergreens as a backdrop, possum haw provides a much needed effect in the gray days of early winter.

Winterberry is not as tolerant of sweet

soils as possum haw but makes up for it with an ability to grow with its roots standing in water as well as sunk in southern red clay once it becomes established. It is also easily the most hardy of the hollies listed here, tolerating −25°. Purplish-green foliage turns black with the first frost. Since winterberry is often found growing along creeks, it is sometimes called black alder. No alder ever had berries like these. A commercial industry for Christmas decoration has been built around cut, fruit-laden winterberry stems.

Numerous winterberry selections have been made, and hybrids with *I. serrata* have been introduced. Too many to evaluate appeared in the late 1980s. 'Winter Red' from Bob Simpson, 'Sparkleberry' from the National Arboretum, and 'Autumn Glow' from Dr. Elwin Orton's breeding program at Rutgers University each has a distinct character of its own as well as producing excellent, abundant fruit that persists as long as birds cooperate.

Itea virginica

Virginia sweet spire, Virginia willow

Habitat: Coastal plain and piedmont wetlands, along streams and in swamps, Florida to New Jersey. Listed elsewhere as hardy to Zone 6, but established specimens have endured colder winters with little damage: probably Zones 5–9.

Propagation: Seeds are tiny. Germinate them under mist or a plastic tent and be patient. Small sweet spire seedlings lack vigor. Semihardwood cuttings root easily with

Figure 57. *Itea virginica* 'Henry's Garnet'.

no hormone treatment any time after tissue firms.

Virginia sweet spire is an excellent, underused plant. White bell-shaped flowers blossom on drooping 3-inch racemes that open from the base to the tip so that the plant appears to bloom for quite a long time. Flowering occurs after the main burst from other spring-flowering shrubs, early May in Georgia but mid-June in Baltimore. Reddish fall foliage persists well into winter and is semideciduous in all but the warmest locations.

This itea is best planted on the edge of a wooded area or as part of a border where it will receive some direct sunlight. It tolerates both moist and dry slightly acid soils, particularly when grown with an organic mulch. Plants in the sun will remain compact at 3–5 feet tall, forming a mound of arching branches, each tipped with a creamy white flower. Plants have more flowers and far better fall color if grown in full sun at least part of the day. In deep shade, plants survive but flowers are sparse and fall color is disappointing. The selection

Figure 58. *Kalmia latifolia*.

'Henry's Garnet', originally found in Georgia by Mary Henry, will have flower spikes up to 6 inches long with fall foliage that ranges from reddish-pink through scarlet to burgundy. This selection was one of the first six plants to receive the Styer Award of Garden Merit after its release by the Scott Arboretum at Swarthmore, Pennsylvania.

Minimal garden care is required. Virginia sweet spire grows and blends well with most other border shrubs but can survive the slightly soggy spots that others cannot. I've never felt the need to prune mine; if you do, prune individual canes all the way to the soil during late winter. If you must fertilize, use the low rates suggested for azaleas and rhododendrons.

Kalmia latifolia
Mountain laurel, calico bush

Habitat: Dry rocky woodlands, outcroppings, slopes, and streambanks from Florida to Canada. Zones 5–9.

Propagation: Seeds are tiny and mature late. We collect the round seed capsules no earlier than November in the southern Blue Ridge Mountains. Germinate seeds under mist or a plastic tent. Cuttings from most wild plants are nearly impossible to root. Serious growers should consult the excellent book by Richard A. Jaynes listed in the References.

Since mountain laurel was first discovered by colonists coming to America, perhaps no other woody native shrub has more excited and frustrated gardeners worldwide. When it is in full bloom, nothing quite compares.

In recent years, research at a number of universities and the breeding work of Dr. Richard A. Jaynes have provided hope and inspiration. Nursery-grown mountain laurels are now available at better garden centers as both wild types and the beautiful hybrids created by Jaynes and others. We're still learning which of the hybrids perform best. Few seem to do well in the Deep South, but 'Carousel', 'Elf', 'Nipmuck', 'Olympic Fire', and 'Sarah' all do well from the cooler parts of Zone 8 north. A white-flowering South Carolina selection, 'Pristine', holds promise as the first named variety specially selected for warmer areas.

Many failures with mountain laurel in the landscape are due to providing too much care after the first year. Mountain laurels are quite tolerant of dry soils but intolerant of wet sites. Research at North Carolina State University has shown dramatic benefits in both survival and growth as a result of mixing a few inches of pine bark or peat into mineral soil before planting and ensuring a few hours of sunlight each day. Also, use a light hand when fertilizing. Mountain laurels thrive on less fertilizer than is required to maintain a healthy rhododendron. Full sun is often too much in Zone 7 and warmer areas, but in cooler areas it should be considered to prevent legginess, limit leaf spot, and encourage flowers. If plants do get leggy, don't be afraid to prune them hard in late winter. You'll lose one season's bloom but increase flowering in following years.

Leucothoe fontanesiana (catesbaei)

Drooping leucothoe, doghobble, fetterbush

Habitat: Moist piedmont and mountain woodlands, often in streamside thickets, from Georgia to Virginia. Zones 6–8.
Propagation: Seeds are tiny. Germinate them under mist or a plastic tent. Semi-hardwood cuttings root easily with .5 percent IBA powder, while hardwood cuttings should be treated with .8 percent IBA powder.

The best garden leucothoes are grown for their foliage rather than flowers. Drooping leucothoe will have dark green glossy foliage year round in the shade. If planted where it receives some sunlight in the winter, foliage will turn bronze to burgundy following the first frost. New growth is often quite red, with the selection 'Scarletta' probably having the reddest. A variegated selection, 'Girard's Rainbow', is becoming widely available.

A coastal species, *L. axillaris*, with similar arching canes of dark green glossy foliage is becoming popular in the nursery trade. Its maximum height is 4 feet instead of the 6 feet achieved under ideal conditions by *L. fontanesiana*. Neither is terrific in the humid Deep South unless given at least 6 hours of sunshine each day, because leaf spot diseases find them so attractive. *Agarista populifolia*—until recently *L. populifolia*, or Florida leucothoe—is a far better plant to use for a similar foliage effect in warmer, humid areas. Florida leucothoe has a medium green color and a more upright habit of growth, approaching a height of 12 feet under ideal conditions. All three do very well

Figure 59. *Leucothoe fontanesiana*.

in full Zone 7 shade, growing in very moist soil. They fill an important void in natural water garden landscaping. In the cooler parts of Zone 7 and all of Zone 6, drooping leucothoe should be the plant of choice; but further south, any of them deserves a try.

Lindera benzoin
Spice bush

Habitat: Moist piedmont and mountain woodlands, often found in streamside thickets and springheads. Spice bush is a good indicator plant for limestone outcroppings from Florida to Canada. Zones 5–9.

Propagation: Remove seeds from fruit; then provide a month of warm stratification followed by at least 3 months of cool stratification. I've never been successful with cuttings, but some people report luck with semihardwood cuttings.

Spice bush is not just another medium to large (to 12 feet tall) shrub that will grow in moist soils and has scarlet berries that attract birds, although that may be reason enough to grow it for wildlife gardeners. Spice bush also has interesting greenish-yellow flowers early in the spring, before leaves emerge, when any hint of color is welcome. Its fall foliage color is a brilliant yellow, even when grown in deep shade.

Figure 60. *Lindera benzoin*.

Light shade, however, yields a bush with better form and more berries.

My chief reason for wanting spice bush in the world is its fragrance. Brush against the foliage or squeeze the fruit, and you'll be treated to a scent reminiscent of many spices but not exactly like anything other than spice bush—a bonus on any woodland stroll. For centuries, the bark has been part of medical treatment for dysentery, coughs, colds, and almost any sort of breathing problems. To be sure your plant thrives, sprinkle the earth with dolomitic limestone annually if your soil is naturally acidic. Spice bush responds to sweet soils better than most other eastern natives.

Liriodendron tulipifera

Tulip tree, tulip magnolia, tulip poplar, yellow poplar

Habitat: Usually found in rich, moist woods, but it will colonize seemingly inhospitable sites in dry, abandoned fields from Florida to New England. Zones 5–8.
Propagation: Seeds generally germinate in low percentages. Commercial plantings almost completely blanket the earth with seeds to obtain a population of 25 plants per square foot. Stratify seeds for 90 days.

No tree provides superior vertical accents in the landscape. However, because tulip poplar's wood is relatively soft and its roots are aggressive, don't plant one closer than 50 feet from your house.

Unfortunately, tulip poplar blossoms are often missed because they are up 50 feet or higher in the tops of trees. That's too bad, because they provide some of the nicer, more dependable mid-spring tall-tree flowers, as well as being a staple of "wildflower" honey production in much of the Southeast. Another feature providing multi-season interest is the fall foliage, which turns a consistent, clear golden-yellow.

Outside its native range, particularly in the Pacific Northwest, this close relative of magnolias is enjoying increased popularity as a street tree. Why? It responds well to soils of average or poorer fertility, growing fairly quickly into a tall, straight, pest-resistant tree that provides dense, cool shade in an urban environment. Once established, tulip poplar responds to drought with prematurely golden foliage that drops to reduce the number of leaves losing water rather than turning brown or dying.

In colonial days "yaller" poplar ranked with chestnut in overall value. It provided the walls of homes, furniture, eating utensils, musical instruments, bark baskets, and sweet honey for the taste buds. Today, the wood is used to frame furniture and provide the core of plywood as well as having many uses in traditional fine crafts. Woodsmen call larger trees yellow poplar and smaller

Figure 61. *Liriodendron tulipifera*.

trees white poplar because of the color of the heartwood. The name "tulip" is supposed to originate from the shape of the flower, but notice the profile of the leaf. Does it look like a tulip?

Magnolia

Mention magnolias and visions of the soft, warm, delightfully scented breezes of the American South fill the mind. Magnolias are an integral part of southern landscape beauty. At the same time, they have served many utilitarian purposes, providing medicines, food for wildlife, and wood for furniture as well as attracting Europeans to the plants of the New World. Sweet bay, *M. virginiana*, was introduced to England in the late seventeenth century, with southern magnolia, *M. grandiflora*, following in the early eighteenth. Until the trees were introduced in Europe, early explorers' descriptions of southern magnolia were not believed. Showy plants of such magnitude were not thought to exist outside tropical climates.

Magnolias are a relatively diverse group, including both evergreen and deciduous species that will tolerate winter temperatures of −20° or lower. Some, like sweet bay, have evergreen strains in the southern part of their range and deciduous ones where winters are colder. Recent selection work is revealing magnolias with apparently much greater cold tolerance, such as southern magnolias that have withstood −25°. This, together with current breeding efforts, gives hope that magnolias may extend their range of garden usefulness even further in the near future.

Native magnolias grow best when roots have plenty of moisture available. However, many will tolerate occasional periods of very dry weather once established. Although they grow best in neither very alkaline nor very acid soils, they do tolerate a wide soil pH range, from 5.0 to 6.8, and still look good. This tolerance for widely differing conditions, as well as their beauty, makes some magnolia species valuable landscape trees.

I have listed and illustrated only one truly deciduous magnolia, M. acuminata. However, there are at least six native deciduous magnolias. Telling them apart is not always easy for the casual observer. M. ashei is a shrub form native only to the Florida panhandle, but it gives horticulturists hope that they can have a large-flowered magnolia with flowers at a height that allows them to be easily seen from ground level. The leaves of Ashe magnolia are very similar to those of bigleaf, M. macrophylla, so named because its leaves are often 1 foot wide and 3 feet long. Pyramid magnolia, M. pyramidata, is a small tree of the Deep South, sometimes located where Ashe magnolia might be found. The underside of Ashe magnolia leaves are white and sometimes hairy, while the undersides of pyramid magnolia leaves are smooth and green. In the mountains

Figure 62. *Magnolia acuminata.*

from Georgia to Maryland, you might encounter Fraser's magnolia, M. *fraseri*. It is fairly common along rivers and streams. The flowers are creamy white and fragrant. The trees are usually small, and the leaves are clustered toward the end of the current season's growth. Umbrella magnolia, M. *tripetala*, also has the characteristic of leaves clustered at the end of stems to resemble an umbrella. The leaves on M. *tripetala* are often long and tapered toward the base.

There are other species and varieties of deciduous magnolias that you might come across in your travels. However, if they have pink or purple flowers, they are not native.

Magnolia acuminata
Cucumber tree

Habitat: Rich woodlands from Georgia to New York, generally in the piedmont and mountains. Zones 5–8.
Propagation: Seeds should be removed from fruit shortly after harvest and handled with care. Excessive drying or heat can

damage seeds. Sow seeds directly in the ground outdoors or cold stratify for 3 to 6 months. Embryo dormancy is variable from tree to tree and season to season, so if seeds don't germinate the first year, wait another year.

Cuttings root with difficulty when they root at all. Little experience exists in rooting cucumber tree (I found no one who had been successful). If you want to try, use summer cuttings but otherwise treat the cuttings as you would those for *M. grandiflora*.

Cucumber tree is named for its unusual red fruit, the shape of which resembles a twisted cucumber. This tree is large in every way; it can grow up to 100 feet tall with a trunk up to 4 feet in diameter. In spring, its leaves appear at the same time as its flowers. The medium green leaves are usually 6–8 inches long. The 3-inch blue-green flower buds open to greenish-yellow bell-shaped flowers in mid-spring.

This is a tree only for large landscapes where the huge leaves dropping throughout the late summer and fall are not a maintenance problem. It is most effective in a landscape where you can look down on the canopy of leaves since the flowers and interesting fruit are often at the top of this potentially massive tree.

The best native deciduous magnolia for average to small home landscapes is the less commonly seen *Magnolia tripetala*. Its 6- to 8-inch-diameter creamy white flowers that bloom in mid-spring can provide an effective highlight against dark woodlands or near water. However, be forewarned that the floral scent of umbrella magnolia is unpleasant, so keep it away from living areas.

Magnolia grandiflora
Southern magnolia, bull bay

Habitat: Moist woodlands and hammocks from Florida to North Carolina. Zones 6–9.
Propagation: Collect cones as they drop or as soon as red seeds (which look like M&M candies) appear. Cones can be dried in paper bags at room temperature until seeds emerge. These seeds are a favorite food for some wildlife, so be alert or you will only find empty cones. Treat seeds like those of *M. acuminata*. Some propagators have reported that germination occurred in two weeks with no stratification for seeds immediately cleaned and sown or soaked in water for a week at room temperature and then sown. Others insist stratification is required. I suggest planting fresh seeds and waiting for them to come up whether it takes a week or months.

Recent research has reported successful rooting of wounded mid-summer cuttings using high concentrations of IBA or NAA, but results from clone to clone have been highly variable. Perfect media drainage is essential for successful rooting. Mike Dirr suggests 100 percent perlite.

I've seen magnificent landscape specimens of southern magnolia on three continents. Few American trees have been so prized in foreign landscapes. In fact, *Magnolia grandiflora* is so coveted in England that gardeners there go to great lengths to bring it into bloom. Because southern magnolias are reluctant to bloom in the cool English climate, they are often grown against heat capturing south-facing walls or trained on these walls as espaliers to bring them into bloom.

Figure 63. *Magnolia grandiflora*.

However, there is little difficulty in growing *M. grandiflora* in the southeastern United States. America's warm summers are superior to those of England for this terrific tree.

Southern magnolia is a relatively fast growing tree once it gets established, but it doesn't get extremely large in most of its range. Bold, large, deliciously fragrant waxy white flowers appear throughout the summer and into early fall. The dark green glossy foliage is prized in floral design. The brown, feltlike fuzz found on the backs of leaves on many trees is sought by floral designers to add long-lasting texture and color to winter bouquets.

Southern magnolia should be given space to develop and then used as a specimen plant. Because it is evergreen and covered with dense foliage, little shade penetrates to the earth beneath it. This lack of light coupled with an aggressive, shallow root system make *M. grandiflora* a difficult plant to blend into a border. However, at the edge of a pond or stream, where it never suffers for lack of water, southern magnolia is magnificent. If you inherit a massive old specimen, don't be afraid to gradually open up the base by pruning and thinning. This has been done effectively at Colonial Williamsburg and other southern restorations.

Magnolia virginiana
Sweet bay, swamp magnolia

Habitat: Wetlands, Florida to New England. Zones 6–9.

Propagation: Cuttings are at least as easy as seeds. Semihardwood cuttings taken from young trees and treated with .8 percent IBA powder or a 2,500 ppm IBA liquid quick dip root readily. Seeds require 2 to 3 months' cold stratification.

Figure 64. *Magnolia virginiana*.

Sweet bay is a highly underused small tree. The spicy fragrance from bruised leaves and bark is attractive. The 3-inch-diameter creamy white flowers are lightly lemon scented, giving them a delightful fragrance less bold than that of southern magnolia.

Sweet bay is at its best in moist locations where the soil is not above pH 6.5. One of its identifying and most charming characteristics is the flash of blue-white to silvery color that can be seen when trees lift their leafy skirts in a breeze. Sweet bay is nearly evergreen in Zones 8 and 9, while it is mostly deciduous further north.

Malus angustifolia
Southern crabapple

Habitat: Thickets at the edge of woodlands or the moist edges of abandoned fields from Florida to Maryland and West Virginia. Zones 5–8.

Propagation: Divide by removing suckers and root sprouts in late winter. Seeds require 60 to 90 days' cold stratification. Removing seeds from fruit and fall sowing is easiest.

Rootability of cuttings varies widely from tree to tree. Softwood or early hardwood cuttings treated with 2,500 ppm IBA solution or .8 percent IBA powder in a well-drained medium have rooted in 10–12 weeks for some commercial varieties, while others treated exactly the same produced no roots at all.

The tree or shrub that looks like an apple along roadsides throughout the eastern United States is probably an apple or crabapple. The early history of European settlers in North America is closely tied to apples; seeds from apples that may have originated almost anywhere in the world have sprouted along the wayside as well as having been sown intentionally since the earliest years of America's colonization. Add to that the hundreds of named varieties of flowering crabapples crossbred by honeybees and you can understand why determining the exact lineage of a roadside apple is difficult.

Determining which apple or crabapple you are seeing is not worth the time for most gardeners. *M. angustifolia* is probably the single most commonly seen true species

Figure 65. *Malus angustifolia*.

in the Southeast, but even more commonly seen are seedlings from diverse sources. Enjoy the fleeting fragrance and beauty of the flowers, be thankful for the fruit that feed deer and other wildlife in the fall, and then go to a garden center or nursery catalog to choose a suitable plant for your yard.

The likelihood of producing a disease-resistant crabapple with desirable flower or fruit characteristics from a randomly selected seed is minute. Therefore, if you are seeking a crabapple for ornamental purposes, please consult your local extension agent or nursery. They can suggest varieties with flowers that look like native crabapples and fruit that is probably better (but just as attractive to wildlife) and will not succumb to the myriad maladies that plague apples in the humid eastern United States. This is one plant where native is not better.

Crabapples should be given ample space, good drainage, and well-limed soil in order to perform best. Fortunately, crabapples are among our toughest competitors once established, so even if you can't provide ideal conditions, you may want to consider planting one. Few trees are as spectacular when in bloom, and few have as little appeal thereafter. Therefore, plant your crabs where they can be appreciated during apple blossom time but in a location near the woods. They should be located away from outdoor recreation so that wildlife can take care of fruit disposal and so that the insects, especially yellow jackets, that are attracted by some varieties in the fall will be well away from living areas. There are no crabapples in my garden, and I helped the neighbors cut theirs down. Crabapples are wonderful trees in someone else's, but not an immediate neighbor's, garden.

Myrica cerifera
Southern wax myrtle

Habitat: Coastal plain and piedmont from Florida to New Jersey. Zones 7–9.
Propagation: Seeds require 60 days' cold stratification. However, nursery growers in Florida say they get good germination of fresh seeds with no stratification. Many seeds have a waxy coat that prevents water uptake and subsequent stratification. Soaking seeds in hot water or rubbing them vigorously against a rough surface will help remove the wax.

Semihardwood cuttings treated with 10,000 ppm IBA solution root well. However, research shows that timing—take cuttings in early summer—is more important than hormone.

Root cuttings 2–3 inches long should be made in early winter. Propagating media should be coarse sand rather than peat or other organic media.

Figure 66. *Myrica cerifera*.

I rarely stroll by a wax myrtle without crushing and smelling a leaf. The spicy fragrance is pleasant and somewhat stimulating. New spring growth exudes the aroma of bayberry candles. Campers tell me that a leaf thrown in the stew pot can almost substitute for bay leaf and that rubbing the skin with crushed fresh foliage will help repel insects. A neighbor in Florida insisted a wax myrtle hedge kept mosquitoes out of her patio. It may have kept some out, but I nonetheless managed to give blood while visiting with her on certain still evenings. Since wax myrtle responds well to pruning and the sandy-soiled suburban Gainesville neighborhood where we were living offered limited privacy, it was the perfect hedge, mosquito-repelling or not.

Wax myrtle is more useful than showy. It will grow in wet sites as well as in some of the driest, excessively drained sandy soils. An added bonus is that wax myrtle has excellent tolerance of salt spray and high soil pH. When provided with occasional irrigation and some fertilizer, it becomes a vigorous background or hedging plant. Wax myrtle's green foliage is also very effective when contrasted against the weathered gray woodwork of many coastal homes.

Obtaining plants from either a local source or a cooler area is important. Great variation in cold tolerance exists from one end of the range to the other. Plants from Florida, for example, have not been hardy during a cold Atlanta winter.

There are two potential drawbacks to wax myrtle. The first is its size. It tends to get leggy when grown in the shade, and becomes too overpoweringly large for most sites when grown in full sun and good soil. Native wax myrtles usually need to be pruned heavily every year or two unless space is unlimited. Fortunately, at least two dwarf wax myrtles have been discovered; nurseries are currently increasing numbers and working on plant patents. The second drawback is winter breakage resulting from ice that accumulates on the evergreen foliage and twigs. Fortunately, wax myrtle recovers rapidly from the hard pruning often required after an ice storm.

Myrica pennsylvanica
Bayberry

Habitat: Coastal wetlands and adjacent areas from North Carolina to Newfoundland. Zones 2–7.

Propagation: Collect seeds during mid- to late fall and process as with those of wax myrtle. They require 90 days' cool stratification. Semihardwood cuttings are reported to root moderately well with a 5,000 ppm IBA liquid quick dip.

Despite being adaptable to a wide range of growing conditions from salty wetlands to dry, infertile sands, bayberries are hard to find in nurseries except those near the sea. If you want them, you may have to grow your own. Like hollies, bayberries have separate male and female plants. Only the females produce berries, but without the males to provide pollen no berries will appear. Therefore, if you are introducing bayberries to your area, you'll need both male and female. Fortunately, if you grow a seedling population, you are almost certain to have both sexes.

The species exhibits a wide range of heat and cold tolerance. Therefore, it is very important to grow or obtain plants from a climatic region similar to your garden. Newfoundland bayberries may not adapt well to the Eastern Shore of Virginia and vice versa.

In southern parts of the range, the fragrant, leathery, gray-green to olive-colored leaves remain on the plant during the bleakest seasons, providing hope during the gray of raw coastal winter days. Further north, bayberry is deciduous. When planted near deciduous hollies, or cut to bring indoors, bayberry helps the spirit endure. The natural waxy coating of seeds has been used as a candlemaker's resource since shortly after European settlers made their way to America's shores. Today, most bayberry candles contain just enough bayberry wax to provide fragrance. The candles burn longer, smell nearly as good, and the limited amount of processed bayberry wax goes further.

Ostrya virginiana
Hop hornbeam, ironwood

Habitat: Found in moist, rich woodlands as well as near springs and along streams from Florida to Nova Scotia. Zones 3–8.

Propagation: Rub the ripe nutlike seeds together and screen them to remove chaff. Seeds are supposed to require 3 to 4 months' warm stratification followed by 60 to 90 days' cool stratification. Despite my problems in germinating seeds, plenty of trees seem to exist along streams in the southern Blue Ridge, so germination must be possible.

Figure 67. *Ostrya virginiana.*

Figure 68. *Oxydendrum arboreum*.

Hop hornbeam is one of the small trees often mentioned when landscape architects are asked which native trees they would like to see introduced into the nursery trade. Once established, it is apparently quite tolerant of the stresses caused by urban living, has no major pests, and will tolerate drier sites if spring transplanted and irrigated the first season.

The "hop" portion of this tree's common name comes from the appearance of its flowers and seeds, which are said to resemble hops used by brewers. Both the unusual flowers and seeds are eaten by birds. The wood of this small to medium-sized tree is very dense (like iron?) and was often used for tool handles by colonists.

Oxydendrum arboreum

Sourwood, sorrel tree, lily of the valley tree

Habitat: Dry hillsides, woodland edges, and road cuts from Florida to Pennsylvania. Zones 5–8.

Propagation: Seeds are tiny and hairlike, ripening well after leaves have dropped, usually in November in the southern Blue Ridge. Treat seeds like those of native azaleas. Softwood cuttings rooted in high percentage when I treated them with .8 percent IBA powder.

Sourwood has been prized for its landscape value as well as its utility since colonial days. Sourwood honey is justly renowned for its flavor. Tool handles and wooden runners for mountain and hill sleds were often constructed of sourwood.

In the landscape, few small trees can compete with the seasonal appeal of sourwood. The lily-of-the-valley-like flowers appear in early summer when little else is blooming. Seed pods remain until well after leaf drop in the fall. Fall color arrives at widely varying times from tree to tree. It is common to see some trees with carmine foliage and others with totally green leaves on the same hillside in August. By October, most sourwood leaves have turned a brilliant to deep red.

Sourwoods are naturally interesting architectural trees rather than classically beautiful ones. Therefore, I suggest planting them where they can be seen from a distance rather than expecting sourwood to be the focus of your patio planting. Initial soil preparation should be similar to that for sourwood's first cousins, the rhododendrons. However, don't plant sourwood in deep shade. Sites at sunny woodland's edge or in full sun are its favorites.

Philadelphus inodorus
Mock orange

Habitat: Rich, well-drained hillsides from Florida to Pennsylvania. Zones 5–8.
Propagation: Seeds require no special treatment. Softwood cuttings treated with .5 percent IBA powder root readily.

Mock orange is often found in restored colonial gardens and abandoned home sites.

Figure 69. *Philadelphus inodorus.*

However, in most areas it has been supplanted in popularity by its sweetly fragrant foreign cousins and hybrids of *P. coronarius, lemoinei, nivalis,* and *virginalis.*

For a week or two each year, in a Victorian landscape or on a large property, mock orange can be an interesting shrub. The rest of the year it is a large blob of leaves or, during winter, twigs. For those who crave fragrance, sweet mock orange might be worth having in a location where evening breezes can bring sweet-scented pleasure in late spring. Fortunately it will tolerate almost any soil conditions and a great deal of neglect as long as it gets at least 50 percent sunshine. This permits the best use of mock orange, which is tucked away to struggle in a deciduous shrub border.

Physocarpus opulifolius
Ninebark

Habitat: Edges of rivers and streams, moist cliffs, Florida to Quebec. Zones 3–8.
Propagation: Seeds require no special treat-

Figure 70. *Physocarpus opulifolius*.

ment. Semihardwood cuttings root readily without hormone treatment.

Ninebark's white to pinkish flower heads fill a landscape flowering void that often follows the spring bloom explosion. Its flowering peak comes at about the same time as that of sweet bay and tulip tree, just as mock orange is finishing. After flower petals fall, a second, more subtle show from clusters of red seed capsules will take you into summer.

This upright, stiffly arching, 8- to 9-foot shrub bears flowers that droop down toward eye level. The common name comes from the plant's exfoliating brown bark, which typically peels and shreds. This interesting characteristic is often hidden by foliage and younger stems in wild plants.

Ninebark will tolerate almost any landscape situation from dry to wet, acid to alkaline soils. However, it is at its best when ample moisture is available. Ninebark was seen more often in earlier landscapes when houses and gardens were larger and the choice of flowering shrubs was smaller.

Figure 71. *Pieris floribunda*.

Pieris floribunda

Mountain andromeda, mountain fetterbush

Habitat: Balds and moist mountain slopes from North Carolina to West Virginia. Zones 4–6.
Propagation: Seeds ripen in October and should be treated like those of rhododendrons. Softwood cuttings treated with .8 percent IBA powder root, but in fairly low percentages.

Mountain andromeda is an underutilized gem of a native shrub. The foliage is a deep forest green all year round. Flower buds form in the fall and are attractively borne erect throughout the winter, turning progressively more creamy white as they come into full bloom in earliest spring. Best of all, mountain andromeda has a neat, compact habit which permits planting under windows, next to walkways, and in other places where more rapid growers are bothersome if not unwelcome.

One reason this plant is far more popular in New England gardens than in the Deep South is that it is very susceptible to the root rot that often arises in warm southern clay soils. Even in well-drained soils, mountain andromeda doesn't thrive in the heat and humidity of the Deep South.

Figure 72. *Pinckneya bracteata*.

However, gardeners from Baltimore north or at higher elevations should seriously consider mountain andromeda. It thrives on low levels of fertilizer, providing its show when little else is in flower. It also appears to be more insect-resistant than its showier Japanese cousin. If you must have a more showy andromeda and like the characteristics of *P. floribunda*, try the hybrid between our native and the Japanese called 'Brouwer's Beauty'. Flower buds are archingly horizontal, halfway between the erect native and drooping oriental forms. Flower buds are red, opening to white and are followed by distinctly yellow-green new foliage, which turns deep green as it matures. 'Brouwer's Beauty' has the desirable compact form of its native parent.

Pinckneya bracteata (pubens)
Fever tree, poinsettia tree

Habitat: Edges of wet coastal plain woods and swamps, Florida to South Carolina. Zones 7–9.

Propagation: In nursery trials, fresh seeds germinated in 10 days at 75° when sown on top of media under mist with no special treatment, so I suspect they would do the same thing inside a plastic tent.

Fever tree was discovered by John Bartram in 1756. It is unusual to the point of unforgettable in bloom, yet is still not widely popular. The rich green foliage contrasts with showy pink bracts that are displayed for months in mid-summer. This small tree, or large shrub if you don't care to "limb up," deserves a place at the back of Deep South borders where it will get a half day or more of sunlight.

Perhaps fever tree is not more popular because of its rather demanding growing conditions. For best growth, soil should hold moisture but be well drained. This may be why it is usually found wild in the organic soils of bayheads. I wish this splendid tree could survive the cold of our mountain winters so I wouldn't have to drive so far to see it in bloom.

Prunus

Mike Dirr states that the genus *Prunus* contains over 400 species and numerous hybrids. The *Manual of the Vascular Flora of the Carolinas* lists 16 native species with many subspecies. Given the prolific nature of native wild cherries and plums, as well as their attractiveness to animals that sow seeds where they fly, crawl, run, and swim, most of these shrubs or trees will be commonly seen somewhere in the eastern United States.

This book is intended for gardeners and those interested in native plants rather than avid botanists. I think most gardeners will be happier with flowering cherries whose parents originated in the Orient and fruiting cherries that have evolved under the watchful eye of pomologists rather than our natives.

Native cherries and plums have a number of dooryard drawbacks that lead me to prefer the results of the plant breeder's art (science?). First of all, they are messy. Throughout most of the eastern United States humidity is high at the time when leaves are supposed to be on the trees, and these plants are subject to many leaf spot diseases caused by fungi that thrive in humid weather. In addition, insects love them. There are insects that attack only the fruit, producing wormy cherries and plums. There are insects that bore into tree trunks, causing die-back and death. And there are insects that defoliate trees. All of these plagues cause leaves and fruit, or what used to be leaves and fruit, to drop into areas where people might like to walk. Even when none of these problems arises, plants in the genus *Prunus* are often a spring or summertime problem following our feathered friends' feeding frenzies. I pull thousands of seedling black cherries from my garden each year thanks to birds who have feasted in a nearby fencerow.

A final reason to avoid some cherries, particularly if you keep livestock, is prussic acid poisoning. All cherries have some potential to cause poisoning if foliage or twigs are eaten by livestock, but black, choke, and fire cherries are most often implicated in this regard.

Nevertheless, the genus *Prunus* produces too many clouds of white flowers and tons of fruit each year to be totally ignored. Some of the showiest plants are natives that etch memories. The low white fog banks of

Figure 73. *Prunus caroliniana.*

flowers on Chickasaw plum, *P. angustifolia*, in the woods between Cross Creek and Macintosh, Florida, let me know that spring had returned during my Gainesville years. I've moved on, but I'm sure they're continuing to delight others each February. Similar visions dot the piedmont Carolinas each March when the thickets of wild plum, *P. americana*, burst into silvery white bloom. Another, very personal *Prunus* memory is of an afternoon spent on a Blue Ridge mountainside farm stuffing myself with a wonderfully sweet purple-black cherry whose identity I'll never know (although I suspect that it was *P. avium*, grown from seedlings collected along a fencerow and planted behind the barn decades earlier). The farmer claimed they were native, and I was too busy gorging myself to question him.

Prunus caroliniana
Carolina cherry laurel

Habitat: Coastal plain and piedmont fencerows and moist woodlands from Florida to North Carolina. Zones 7–10.

Propagation: Seeds sown in the fall germinate readily the following spring. Some cold stratification is required in the Carolinas, but seeds sown from plants growing near

Clearwater, Florida, germinated in a few weeks with no cold treatment. If there was any stratification, it was minimal. Softwood cuttings root in fair numbers when treated with .5 percent IBA powder.

The greatest use of Carolina cherry laurel is for providing a nearly carefree, dark glossy green visual screen. It is evergreen, flowers briefly in late winter when the bees can use the pollen, produces a small black fruit, and then just sits there the rest of the year. It takes well to shearing, producing the fragrance of Jergens lotion when the leaves are cut or crushed. If allowed to grow, pruned to one trunk, and limbed up, cherry laurel can become an attractive small tree that produces cool, dense shade in hot spots.

While tolerant of some shade, cherry laurel under live oaks or in other dense shade won't form hedges as thick as it would elsewhere. A skeletonizer insect may weave some leaves together and devour most of them, but I've never seen the problem bad enough to warrant spraying.

Prunus pennsylvanica
Fire cherry, pin cherry

Habitat: Mountain balds, following fires and land clearing, from Georgia to Labrador. Zones 3–7.
Propagation: Seeds require 90 days' cold stratification.

This is the cherry seen flowering during spring drives along the higher elevations of the Blue Ridge Parkway. It grows very

Figure 74. *Prunus pennsylvanica*.

rapidly but serves only as a pioneer species, usually disappearing within 20 years unless sites are repeatedly disturbed.

The chief functions of this species are to hold the earth together following environmental disasters and to feed diverse forms of wildlife. Dozens of species of birds as well as bears and smaller animals are known to be fond of fire cherry.

Prunus serotina
Black cherry

Habitat: Old fields, fencerows, and pastures from Florida to Nova Scotia. Zones 2–8.
Propagation: Seeds vary greatly in their stratification requirements from one end of the range to the other. Sixty days' cool stratification is average.

Black cherry has an important part in our colonial heritage. The wood makes beautiful furniture, bowls, and utensils. It is sometimes mistaken for mahogany in color and grain, but you can see the pores in ma-

Figure 75. *Prunus serotina*.

hogany and not in colonial black cherry. Perhaps just as important was its use as a flavoring agent in jellies, wine, and "cherry bounce," a beverage made from black cherries with the aid of rum or brandy.

This is another cherry whose major claims to fame are holding the earth together and feeding myriad wildlife. Perhaps the most universal identifying characteristic of this mid-spring-flowering tree is the webbing of eastern tent caterpillars. These critters will attack other species, but seem to prefer black cherries.

Black cherry can get to be a large tree but isn't one I'd want near my house. It drops lots of leaves, twigs, and fruit. In addition, the reason it is no longer used as a rootstock for grafting more desirable ornamental and fruiting cherries is that borers often discover trees before they acquire much age.

Rhamnus caroliniana
Carolina buckthorn, Indian cherry

Habitat: Moist coastal plain and piedmont woodlands from Florida to Virginia. Zones 6–9.

Propagation: Seeds require 60 days' stratification. I've never tried rooting this plant, but other buckthorns can be rooted using .8 percent IBA powder and softwood cuttings.

Figure 76. *Rhamnus caroliniana*.

Carolina buckthorn's fruit catches the eye first. The slightly sweet fruits on these large shrubs or small trees turn from red to black as they mature. This, together with the glossy green foliage, makes it a tempting plant to include in a naturalistic landscape.

Be forewarned, however. Seedlings near Carolina buckthorn are just as common as cherry seedlings near a black cherry. Therefore, even though an individual plant may be attractive and noninvasive, if you choose to plant one in your garden, you may end up with a weed problem or Carolina buckthorn transplants for all your friends.

Rhododendron

Despite differences in appearance, our native rhododendrons have several characteristics in common. (1) With only a few exceptions, they must have excellent soil drainage. If roots stay wet, they rot and die. (2) Soil must be acidic. A soil pH between 4.8 and 5.5 is ideal as long as other essential nutrients are present. (3) They are light feeders. More rhododendrons suffer damage from the application of too much fertilizer than starve due to lack of fertilizer. Use one-half the suggested landscape maintenance rate shown in Appendix 1 the first year and then no more than the suggested maintenance rate in following years. Rhododendrons reward patience and moderation. (4) The more sunlight they get, the more flowers you get. However, in most of the mid-Atlantic and southern states rhododendrons are at their best in light shade or where they are protected from afternoon sun on the hottest days. (5) Even native rhododendrons are subject to numerous potential pests, but they are worth the trouble. Few plants can compete with the spectacular beauty of rhododendrons in bloom or the rich, lush foliage of the evergreen types year round.

Evergreen Rhododendrons

All of the evergreen rhododendrons listed below are susceptible to root rot and bloom best when given at least a few hours of sunlight each day. They are also all susceptible to a wide range of pests, most of which are a nuisance rather than being lethal. Proper soil preparation and siting and minimal care in getting plants established will limit pest problems and reward you for decades.

If your mature rhododendrons need pruning, prune them immediately after they finish flowering. They respond to winter pruning beautifully, but this year's flower buds were formed last July. Therefore, any pruning after early August reduces the number of flowers you'll enjoy the next year. Young plants can be encouraged to branch

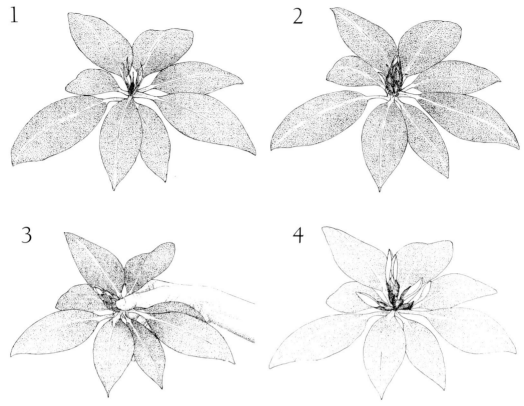

Figure 77. Pruning a rhododendron.

close to the ground, forming dense shrubs, by rolling out the terminal buds of young evergreen rhododendrons just as they elongate to resemble a rocket nose cone or funnel. Once these buds are removed, two to four side branches will usually develop per stem. (See Figure 77.)

Rhododendron carolinianum

Carolina rhododendron, deer-tongue laurel, punctatum

Habitat: Streambanks, rock outcroppings, and mountain balds in the southern Blue Ridge and Smoky Mountains. Zones 5–7.

Propagation: Seeds are tiny. Germinate them under mist or a plastic tent. Semi-hardwood cuttings taken in late summer should be wounded and treated with .5–.8 percent IBA powder or 2,500 ppm IBA liquid. Rooting requires 6 to 8 weeks.

Native rhododendron names are confused. Some people say *R. carolinianum* is the same as *R. minus*, while others call it *R. minus var. carolinianum*. One respected authority suggests we use the common name and let the botanists argue about the scientific name. I agree. The horticultural differences between these plants are real whether the botanical differences are important or not.

The plant I call Carolina rhododendron is

Figure 78. *Rhododendron carolinianum*.

one of the hardiest native evergreen rhododendrons grown. When grown in full sun, it is a dense 4- to 5-foot mound of evergreen foliage, some leaves turning mahogany red in the winter. When the foliage is crushed, there is a distinct fragrance of aromatic herb oils.

In early spring, before dogwoods bloom and before new rhododendron leaves start to grow, perfect 3-inch flower trusses cover plants. The most common color is pale lilac-rose, but some plants have pastel pink or pure white flowers. When grown in warmer areas, a few flowers may open in late October or November.

Rhododendron catawbiense
Red or purple laurel, mountain rosebay, Catawba rhododendron

Habitat: Mountaintops, bluffs, and cliffs from Alabama to West Virginia. Prevalent, forming dense thickets, along higher elevations of the Blue Ridge Parkway, dominating some balds. Widely scattered populations also exist in the piedmont of Alabama, Georgia, and North Carolina. Zones 4–8.
Propagation: Seeds are tiny. Germinate them under mist or a plastic tent, with seeds on top of media so they are exposed to light. Semihardwood or dormant cuttings taken in early winter are possible to root

Figure 79. *Rhododendron catawbiense*.

but difficult. Wound cuttings and treat with 5,000–10,000 ppm IBA solution. They will root in 8 to 16 weeks, if they are going to root at all.

Five- to 6-inch trusses of raspberry-sherbet-colored to pale lavender-pink flowers appear on this plant after the main explosion of azalea blooms (late April in Atlanta, Memorial Day in Philadelphia). White forms are rare but do exist. Festivals are scheduled to view the splendor of acres of natural rhododendron gardens in the southern mountains. In the home landscape the effect of this neat, large-leaved shrub in full flower is rich and lush while orderly. If grown in deep shade, plants tend to become leggy and fail to bloom. In light shade, or sunshine where they can tolerate it, *R. catawbiense* rivals hybrids for spectacular bloom.

This species has been one of the parents of many, many hybrid varieties. Most of these hybrids, like the species, will form a mound 8–10 feet tall and equally wide at maturity. *R. catawbiense* selections currently available from nurseries are better suited for cooler parts of Zone 8 and throughout Zones 7 and 6. However, selection and evaluation is under way from native populations in warmer areas. Hopefully, this work will result in a good Zone 8 or 9 *R. catawbiense* for those few places with adequately drained soil in the hotter parts of our region.

Figure 80. *Rhododendron maximum*.

Rhododendron maximum
Great or white laurel, rosebay, max

Habitat: Along streams, on moist slopes, and in woodlands from Georgia to Canada. Rosebay is the state flower of West Virginia. Zones 5–8.
Propagation: Same as for *R. catawbiense*.

Max has the largest, darkest green leaves of all the native rhododendrons. It also tolerates the deepest shade but does exceedingly well in full sun in cooler parts of its range.

The greatest use of max should be for screening in shady areas where space for this big shrub or small tree is available. The deep forest green foliage with leaves commonly 9–10 inches long provides excellent privacy without the formality or labor of a clipped hedge.

Late spring or early summer white flowers, often suffused with blush pink in cooler areas, are a bonus not found in most screening plants or shade-tolerant shrubs. My 10-by-10-foot plant displayed hundreds of 5-inch white flower trusses in eastern white pine shade this spring. I've seen max in peak bloom during the same week in the southern Blue Ridge Mountains and the Berkshire Hills of Massachusetts.

If you want an evergreen tree rhododendron, max is the most likely candidate among the natives. Specimens 30 feet tall are not uncommon in rich lower mountain coves. Where conditions favor *R. maximum*, walls of dark green climb north- and east-facing slopes above streams.

Rhododendron minus
Minus, June punctatum

Habitat: Streambanks, rocky outcroppings, and wooded slopes of the Carolinas, Tennessee, and Georgia. Found in both the mountains and upper piedmont. Zones 5–8.
Propagation: Same as for *R. carolinianum*.

Figure 81. *Rhododendron minus*.

This plant is very similar to *R. carolinianum*. The chief garden difference is that *R. minus* flowers fully two months later. As a result, flowers are often seen peeking out from under new leaves and stems that have just reached full expansion. This plant is particularly showy around Father's Day each year, hanging from cliffs along roadsides in Transylvania County, North Carolina. Viewed this way, looking up from under the plant, flowers are quite apparent despite new foliage.

When leaves of minus are crushed, there is no fragrance of aromatic oil as there is with *R. carolinianum*, and minus's flowers are usually darker and much richer rose-purple than those of Carolina rhododendron. Another difference is in susceptibility to leaf spot disease. When minus is drought stressed or grown in deep shade, leaf spot is much more severe than on Carolina rhododendrons grown under the same conditions. To me, Carolina rhododendron is the superior landscape plant of these two.

Deciduous Azaleas

Botanically, all azaleas are rhododendrons. All native azaleas are deciduous. Distinguishing among native azaleas can be a challenge, even for experts. One reason is that most species interbreed. Another is that pink- and white-flowering forms abound. Yet another is that humans have moved them out of their native habitat to new homes where they have flourished. If you are confused concerning the identity of a native azalea, perhaps the first thing you should look for is evidence of an abandoned garden or home. The plant you are viewing may not be native to the site at all.

The easiest way to tell native deciduous azaleas apart is to ask yourself the questions listed in the Introduction, remembering that sequence of bloom remains the same even if time of bloom varies throughout the East.

The first native azalea to bloom depends upon where you are. Pinxterbloom, *R. periclymenoides* (formerly *R. nudiflorum*), is often the first in much of the piedmont, with flowers in shades from pink to white. As you move to warmer areas, pinxterbloom is not as commonly found. There, the subtle perfume of the pink to white piedmont azalea, *R. canescens*, may draw you as early as February in Florida and southern Georgia. Those lucky enough to live in the upper mountains of North Carolina will often see what appears to be a flock of pink butterflies in the deciduous forests as oak buds are swelling. Closer investigation will reveal pinkshell azalea, *R. vaseyi*, leafless but covered with flowers. If you are lucky, real butterflies will be visiting the flowers, one of the few plants in bloom this early. I am pleased to say that I had nothing to do with naming these plants, as there is a white form of pinkshell as well as a white pinxterbloom. At least they aren't called whiteshell or whitesterbloom.

By mid-spring, the bloom sequence gets jumbled. Everything seems to flower at once. Flame azaleas, *R. calendulaceum*, usually in shades of yellow and orange, illuminate woodlands from Georgia to Pennsylvania. However, in the Blue Ridge Mountains above 3,000 feet and on the Cumberland Plateau a deeper red-orange azalea with flowers only slightly smaller than those of the flame azalea may be found blooming from late May on through July. This is the Cumberland or Baker's azalea, *R. bakeri*. To further confuse identifiers, a pink-flowered azalea for mid-spring also exists in the mountains from North Carolina to New England. The roseshell azalea, *R. roseum*, can be tall, 15 feet or more, and has clove-scented pink flowers.

Mid-spring is also blossoming time for the white-flowered azaleas. The low-growing, sweetly fragrant dwarf or coastal azalea, *R. atlanticum*, is found in pine barrens and moist woods but is rarely seen in the upcountry. The white, sweetly fragrant swamp azalea, *R. viscosum*, most commonly found in piedmont and mountain wetlands from Florida to New England, is sticky to the touch like the coastal azalea but can reach heights of 10 to 15 feet, while the coastal azalea is usually only 2 feet tall. Swamp azaleas' fragrance is often spicy sweet rather than just sugary sweet like that of coastal azalea. Both are highly stoloniferous, so when you see one plant, you usually see a few more growing nearby. The last of the commonly seen white-flowered azaleas is the smooth or sweet azalea, *R. arborescens*. It can also be found in wetlands but is more commonly seen in the upcountry. By late May, the heliotrope fragrance of the sweet azalea can usually be found throughout the mountains, with the result that the plant is often smelled before it is seen on hiking trails. If you can't find the source of the fragrance while hiking, look up. I've seen sweet azaleas 20 feet tall. Usually, this delightful large azalea will have red inner flower parts with white petals. Both swamp and sweet azalea bloom well into June from the higher southern mountains north to New England.

How do you sort all this out? If your azalea is yellow-orange, it's probably a flame unless you are above 3,000 feet in the Blue Ridge or on the Cumberland Plateau. Then you're probably seeing *R. bakeri*. If your azalea is pink with little fragrance, it's probably pinkshell in the mountains of North Carolina and pinxterbloom almost everywhere else in the East. If it's fragrant, it's probably roseshell (clove) in the upper mountains and piedmont (delicate sweet) in the piedmont or coastal plain. If the site is wet and the petals are sticky, you may have a pink form of the coastal or swamp azalea. None of these is a firm rule, however. They are just guidelines for guessing. Remember that almost every wild azalea you see has grown from seeds and you weren't introduced to the parents. Almost anything is possible, and I've mentioned only the nine most common native deciduous azaleas. There are others.

Most evergreen azaleas should never be planted where their roots remain wet for even a day. However, some of our native deciduous azaleas thrive in damp soils. If you have a moist site, pinkshell, sweet, coastal, and swamp azaleas all may have a place in your landscape, giving you a combined bloom period of two months or more.

To get abundant bloom and healthy, vigorous vegetative growth, prepare soils thoroughly and choose your site well before planting. In most of the eastern United States, azaleas will benefit from phosphate mixed thoroughly with the top 6 inches of soil. As with evergreen rhododendrons, the more sunshine, the more flowers, with the best plants in light shade or full sun only part of the day.

Young deciduous azaleas should be "hard pinched" (remove a couple of inches of new growth) during the growing season or pruned hard during the winter for most species. Rabbits may do this for you the first couple of years plants are in transplant beds or in the landscape. Pruning is the only practical way to get dense bushy growth rather than leggy growth on taller-growing species.

While some species are easier to propagate than others, all these azaleas are propagated in the same way (and no specifics will be listed under the individual plants below). Seeds are tiny, so they should be germinated under mist or a plastic tent. Softwood cuttings should be taken as soon as you notice terminal growth has stopped. Rooting percentage and number of roots can be greatly enhanced by treatment with .5–.8 percent IBA powder or 1,000–2,500 ppm IBA solution. New growth must be forced under lights or rooted cuttings left undisturbed until they have gone through normal winter chilling and new growth starts the following spring.

Rhododendron arborescens
Sweet azalea, smooth azalea

Habitat: Mountain and piedmont swamps, streambanks, and moist woods from Georgia to New York. Zones 5–8.
Propagation: See the last paragraph of the general comments on deciduous azaleas above.

Sweet azalea is tall and fast growing for a native azalea. It is suitable for moist landscapes in woodlands and along streams. In moist shade it may get to be 20 feet tall, but most garden specimens are 8–10 feet tall unless controlled by pruning. The greatest flower display is in late spring after the bright green leaves are fully expanded, but sweet azalea continues to bloom sporadically all summer. The white flowers usually have red reproductive parts and a prominent, sweet, heliotropelike fragrance.

Figure 82. *Rhododendron arborescens*.

For all sweet azalea's size, its flowers are more subtle than showy.

Sweet azalea is a consistent performer where it gets adequate moisture and a few hours of sunlight each day. The fragrant flowers are best during late spring when we choose to linger outdoors in the evening. For most people, however, one sweet azalea per yard or garden provides fragrance enough. More can be overpowering.

Rhododendron atlanticum

Coastal azalea, dwarf azalea

Habitat: Coastal plain pine barrens and wooded wetlands from Georgia to Delaware. Zones 6–9.

Propagation: See the last paragraph of the general comments on deciduous azaleas above.

Gardeners who can use a low-growing azalea with abundant white blooms from mid-April to autumn should consider the coastal azalea. The major flush of bloom occurs just before or at the same time as new leaves begin to unfold. Like the sweet

Figure 83. *Rhododendron atlanticum*.

azalea, it is quite tolerant of soil moisture, but its fragrance is not as overpowering as sweet azalea's. Foliage ranges from a crisp medium green to powdery blue-green.

If grown where it receives uniform moisture, coastal azalea should be given room. This stoloniferous plant can create its own azalea colony.

Rhododendron calendulaceum
Flame azalea, yellow honeysuckle

Habitat: Dry, open mountain and upper piedmont woods, occasionally along streams, Georgia to Pennsylvania. It is often found on west- or southwest-facing slopes. Zones 6–8.

Propagation: See the last paragraph of the general comments on deciduous azaleas above.

Flame azalea was described by colonial plant explorer John Bartram as "the most gay and brilliant-flowered shrub yet known." Flower colors range from pale yellow to deep red-orange, with every variation and combination in between. The unscented flowers are larger and more showy than those of other native azaleas.

Flame azaleas have not been fully appreciated because they have not received the care of hybrids. When given at least a few

Figure 84. *Rhododendron calendulaceum.*

hours of sunlight each day, proper preplanting soil preparation, and pruning when young, a plant in bloom will stop traffic. Anyone with a fondness for bright colors and the courage to fight fashionable purple and gray gardening trends should plant flame azaleas prominently. Even on cloudy days, flame azalea flowers seem to glow from within.

Rhododendron canescens
Piedmont azalea, hoary azalea, Florida pinxter

Habitat: Streamsides and moist areas of the piedmont and coastal plain from Florida to Virginia. Zones 7–9.
Propagation: See the last paragraph of the general comments on deciduous azaleas above.

Piedmont azalea is the most common native azalea in the Southeast, frequently hybridizing with at least four other native species. I've seen them flowering in February in Florida and in October following abnormal

Figure 85. *Rhododendron canescens.*

weather in South Carolina. This sometimes stoloniferous species will form colonies, with one as large as 10 acres having been reported! Add to this its ultimate height of 15 feet and you should prepare to give it room or prune it back every year or two.

The delicate, sweet fragrance of piedmont azalea is among my favorite native shrub scents. Colors range from white to deep rose-pink and, despite its southern origins, it is hardy to at least Zone 5. Superior fragrant pink selections are just reaching the marketplace from the Mobile area. This is the perfect azalea to naturalize in open, moist, deciduous woods along trails planted with early spring wildflowers. The ephemeral fragrance and soft pastel colors can't fail to lift the spirit on an early spring stroll.

Rhododendron periclymenoides (nudiflorum)
Pinxterbloom, wild honeysuckle

Habitat: Lower mountain, piedmont, and occasionally coastal plain streambanks and open woodlands from Georgia to Maine. Zones 6–8.

Propagation: See the last paragraph of the general comments on deciduous azaleas above.

Figure 86. Rhododendron periclymenoides.

The name has nothing to do with the color, which, coincidentally, is usually pink but can be white. "Pinxter" is from the Dutch "pinksteren," which when traced back through its ecclesiastical etymology, translates as the rhododendron that blooms on Pentecost, or 50 days after Easter. Obviously, it was named far north of most of the area with which we are concerned. In the Southeast, pinxter commonly blooms on Easter.

Pinxterbloom is often listed as sweetly fragrant and 8 to 10 feet tall, but the ones I've seen have little or no fragrance and are rarely more than 6 feet tall. One giveaway in identifying this plant in flower is that the petals seem to have a ridge running up their back and as the flower heads mature, petals will fall but hang on the stamens, creating a messy appearance. While this is not my favorite pink native azalea, a blooming specimen placed where it catches the morning sun will make anyone look twice. At least a half day of sunshine seems necessary to keep pinxter from getting leggy.

Rhododendron roseum

Roseshell azalea, election pink, clove azalea

Habitat: The higher mountains of North Carolina north to New Hampshire in deciduous woods and on streambanks. Zones 4–7.

Propagation: See the last paragraph of the general comments on deciduous azaleas above.

Along with pinkshell, roseshell has the purest pink flowers of the native azaleas. However, it also has the most distinctive fragrance, a sugary clove scent. While many other native azaleas move into warmer areas with ease, I've seen few plants of this species in the South. However, a few clove-scented deciduous azalea hybrids are making their way south, so perhaps the children of this native will have more impact than the parent.

Rhododendron vaseyi

Pinkshell azalea

Habitat: Mountain bogs, upland oak forests, and the edges of spruce forests in North Carolina. Zones 5–8.

Propagation: See the last paragraph of the general comments on deciduous azaleas above.

Why list a plant that is native only to the North Carolina mountains? Two reasons: first, when pinkshell is in bloom, it dominates the ridgetop landscape as strongly as

Figure 87. *Rhododendron vaseyi.*

does *R. catawbiense* a month later; and second, pinkshell is one of the finest early-flowering azaleas, for moist as well as moderately dry landscapes, from the Carolinas to New England. It will hold its own in northern gardens with just as many flowers as showy hybrids, more pleasing color than many, and often with both superior cold tolerance and dependability of bloom. When provided with high pine shade and mulch, pinkshell also moves nicely into Zone 8.

The flowers are distinctive in that they appear before new leaves show either in the forest or on this azalea. Flowers lack the tube of most native azaleas, so petals seem separate. Flower color ranges from a clear apple blossom pink to pure white with a yellow throat. Fall foliage turns a deep, rich burgundy, contrasting well with the yellow fall foliage so common on many other native shrubs.

Pinkshell is worth having for the flower display if for no other reason. Most plants I've seen in the landscape have been 5–6 feet in height, but pinkshell has the potential to grow over 15 feet tall. If you collect

Figure 88. *Rhododendron viscosum*.

seeds from pinkshell, don't worry about hybrids with other species. Plant breeders tell me that pinkshell won't cross with anything but pinkshell.

Rhododendron viscosum

Swamp azalea, swamp honeysuckle

Habitat: Mountain, piedmont, and coastal plain swamps and wetlands from Florida to Maine. Zones 5–9.

Propagation: See the last paragraph of the general comments on deciduous azaleas above.

Swamp azalea is very similar to coastal azalea but is usually taller and has a spicier fragrance. Flowers appear after leaves and are usually both white and sticky. A pink form has been reported.

This is a well-behaved, orderly shrub in the landscape despite being stoloniferous in the wild. Mature plants will average about 5 feet tall and at least as wide. A welcome feature for me is that it blooms during that shrub-bloom lull around the first day of summer, about the same time as *R. maximum*. My plant is the blue-green leaf form selected by Polly Hill and named 'Delaware Blue'.

Figure 89. *Rhus glabra.*

Rhus glabra

Smooth sumac

Habitat: Dry roadsides and old fields as well as along railroad trestles and fencerows, Florida to Maine. Zones 3–8.

Propagation: Seeds require scarification. Sumacs are most commonly propagated by early winter root cuttings placed in flats of moist sand. Treating root cuttings with .3 percent IBA powder and then "curing" them for two weeks at 65–70° before planting them in rooting flats increases success. When planting out rooted cuttings in early spring, be sure tops are slightly above the ground and soil is firmed to prevent drying.

This large (to 15 feet) colony-forming shrub is one of our most valuable and spectacular accents in dry waste areas such as impossible slopes where even junipers struggle. The dense heads of greenish-white flowers are transformed into velvety red fruit in late summer. Shortly thereafter, leaves turn to a kaleidoscope of colors, from brilliant red-orange to scarlet and red-violet shades. In the late afternoon sunlight of an early autumn day, no plant has more colorful or spectacular foliage.

Staghorn sumac, *Rhus typhinia*, is a similar but larger sumac that can be found in the mountains from Georgia to Maine. The major differences between the two are fuzzy young stems and medium green foliage on staghorn sumac as compared to the dark green leaves of smooth sumac. Staghorn sumac will occasionally turn yellow-orange in the fall, while smooth sumac is almost always dark red. Both have exotic-looking cut-leaf forms called 'Laciniata' that deserve wider use in American landscapes.

Hal Borland wrote that "if sumac were less common here, if it demanded care and pampering, it probably would be cherished and admired simply for its own beauty." I agree. Unfortunately, smooth sumac is all too common and has been too often confused with a skin-irritating cousin of poison ivy. Both smooth and staghorn sumacs were used for medicines by Native Americans. Sumac bark, leaves, and roots were all used for dyeing and tanning by European colonists.

The berries can be used to make a lemonadelike drink. To concoct sumac-ade, June T. Smith advises, begin by collecting the scarlet fruit, rinsing it off, and pulling the fruit from the seeds until you have enough to fill a blender without pushing down hard. Fill the blender with water to completely cover fruit and puree for several minutes. Strain the liquid through several layers of cloth and sweeten to taste. This drink is best consumed at dusk on a hot, humid late summer's day.

Sumacs do best when planted in full sun in well-drained soils. They tolerate neither shade nor wet feet for long. While insects of many types like to feed on sumacs, I've never seen them do any major damage to plants properly sited. Clumps are best rejuvenated every few years by cutting them to the ground in mid-winter. Because these clumps are usually formed by suckering from a single plant, they are often all the same sex. You get berries only on the female plants and you get the fuzzy terminal branches only on males. These naked, fuzzy terminal branches look a bit like velvety antlers and give staghorn sumac its name.

Robinia hispida
Bristly locust

Habitat: Dry roadside slopes and open woodlands from Florida to West Virginia. Zones 5–8.
Propagation: Seeds require scarification. Spring root cuttings are easy if cuttings are stored cool (40–50°) in nearly dry sand to callus for about 3 weeks before planting.

Figure 90. *Robinia hispida*.

If you see a shrub or thicket with pink to bright rose pealike flowers in mid-spring, you are probably seeing bristly locust. The abundance of plants with deep rose-colored flowers on steep sandy slopes throughout the South may be due to the Soil Conservation Services introduction of a bristly locust selection named 'Arnot'.

The chief landscape value of bristly locust is more functional than ornamental. Its

Figure 91. *Robinia pseudoacacia*.

flowers last a short time, but its ability to form dense thickets that help stabilize dry sandy areas and prevent soil erosion continues year round. As a legume, it also fixes nitrogen from the air, so it can survive soils that are almost nutritionally sterile.

Robinia pseudoacacia
Black, yellow, or white locust

Habitat: Mountains and piedmont, usually on disturbed sites, Georgia to Maine. Zones 3–8.
Propagation: Techniques are the same as for *R. hispida*.

A black locust tree covered with pendulous white wisterialike flowers can be breathtaking. The floral fragrance is intoxicating and transforms into a delicious honey that many mountaineers prefer to sourwood honey. Another virtue is hard, long-lasting wood. Folk wisdom suggests that you pound locust fence posts into the earth with a rock and then set the rock by the post. When the rock "rots," you'll need to replace the fence post. Locust posts may not be quite that durable, but they're my first choice.

However, in the landscape, the flaws of black locusts vastly outweigh their virtues. They should remain along road cuts and in disturbed areas where they function as a pioneer species, holding the earth together, adding nitrogen to the soil, and being replaced by forests within a generation. Young growth is covered with vicious thorns, root sprouts will penetrate patios, branches fall in the slightest breeze, and bright yellow foliage starts dropping on the lawn with the first dry weather of summer, often leaving trees naked by Labor Day.

Rosa

More than one hundred species of native roses have been described for the United States. Many possess desirable traits: (1) they have attractive flowers, even if the flowers are single instead of the double tea rose form better known by the general public; (2) they often smell like roses; (3) even when they do become infected with black spot or mildew they continue to grow vigorously; and (4) they have excellent tolerance for salt or drought or flooding, depending upon the species. Unfortunately, they all possess one major undesirable characteristic in that they bloom for a relatively short time and then seem to disappear.

Before you say, "Why didn't Dick include the Cherokee rose or dog rose or multiflora rose or *Rosa rugosa*?" let me inform you that they are all introduced species. Even though they have become naturalized in many areas, they are not native to the United States. In fact, the rose that visitors mistake for a native in my garden is *Rosa rugosa*, a most remarkable plant with legendary salt spray and cold tolerance as well as leathery dark green foliage, large fragrant flowers, and attractive red fruit (hips). But this is a book about native eastern U.S. plants, so none of these roses is included.

Rosa carolina
Carolina rose, pasture rose

Habitat: Dry pastures and edges of dry woodlands from Florida to Nova Scotia. Zones 4–8.
Propagation: Seeds require 90 days' stratification. Semihardwood cuttings root readily when treated with .8 percent IBA powder.

Figure 92. *Rosa carolina*.

Carolina rose flowers are most often a delicate rose-pink, but a white form also exists. The delightful if subtle fragrance is what most of us consider truly rose scented. It rarely grows to be more than 3 feet tall, but it suckers readily, so thickets are commonplace. Carolina rose prefers full sun and well-drained soil. It does well in stony soil and along dry roadsides.

A wild rose, *Rosa palustris*, with similar, if somewhat darker, rose-pink flowers, is found in low, wet areas. As a result, it is called swamp rose. Besides the difference in habitat preference, swamp rose also differs

Figure 93. *Rosa virginiana*.

from Carolina rose in size, growing to be 6–8 feet tall, or about twice as tall as Carolina rose.

Rosa virginiana
Virginia rose

Habitat: Moist areas as well as sandy soils, particularly in the coastal plain, from the Carolinas to the Canadian Maritimes. Zones 3–8.
Propagation: Same as for *R. carolina*.

In addition to having the single pink flowers common to other native roses, Virginia rose bears fruit that is shiny red and persists during the winter. Its foliage is an excellent glossy dark green in summer, changing from purplish to red-orange to yellow in the fall.

Of the native roses, *R. virginiana* is closest to the near-perfect introduced seaside rose, *R. rugosa*, in its tolerance for both salt spray and the drought stresses imposed by excessively drained sandy shore soils. Flowers often appear in clusters of two or three, and a white form exists. It responds well to severe pruning, forming an effective hedge. Even if unpruned, however, plants rarely exceed 6 feet in height.

Figure 94. *Rubus argutus.*

Rubus argutus
Highbush blackberry

Habitat: Fencerows, roadsides, and abandoned pastures from Georgia to Nova Scotia. Zones 4–7.

Propagation: Cuttings are treated like those of *R. odoratus* (see below). However, divisions, tip layering, and digging up suckers are the most common methods of propagation used by fruit gardeners.

Blackberries, dewberries, and raspberries all belong to the complex genus *Rubus*. None, I feel, has a place in the ornamental garden, but any of them might be squeezed in along a fence in a natural or nearly wild landscape.

If you have enough space, and the right palate, you may want to include some in your garden of edibles. Fresh bramble fruit is a summer taste treat to be cherished. Before choosing a site and committing yourself to the work necessary to produce a crop of berries, please consult your local county agent to learn not only which ones to grow but also how to grow them.

In the southern mountains and much of the upper piedmont, the season of peak blackberry bloom is a time of anxiety. It is often referred to as "blackberry winter" because the time of the mid-spring full moon is supposed to be when late frosts claim

Figure 95. *Rubus odoratus*.

newly emerged beans and freshly set-out tomatoes. For me, blackberry winter is usually in May. Old-timers who have been burned once too often seem to plant enough garden for an early harvest every year but wait until the new moon following blackberry winter to set out the main crop of tomatoes and sow beans. If your garden is in a frost pocket, maybe you should watch for blackberry bloom as well.

Rubus odoratus
Flowering raspberry

Habitat: Fertile soils, often the moist edges of hardwood forests, from Georgia to Quebec. Zones 4–6.

Propagation: I've never tried to germinate flowering raspberry seeds. One propagator tells me flowering raspberries can be rooted from softwood cuttings but will rot if they are mature enough to have pith in the stem. The surest way to get a plant is by digging up suckers or getting a division from a friend.

The value of this raspberry is not in its fruit but in its flower. The fruit is nearly tasteless. Some designers might say the same for the flowers. They are at least 2 inches across, opening as a deep purplish-pink and then fading to magenta-pink with age—not an easy color to match with other flowers in the landscape, but dramatic against conifers or mixed with ferns.

However, if you have a spot of good soil with ample moisture, try flowering raspberry as a background plant. The 5- to 6-foot thornless canes are covered with rich green leaves even before the flowers begin to show in early summer. *R. odoratus* provides luxuriant, almost tropical, texture with little care.

Salix discolor
Pussy willow

Figure 96. *Salix discolor*.

Habitat: Moist soils along the edges of streams, swamps, and moist meadows, North Carolina to Maine. Zones 5–8.
Propagation: Pussy willows must produce seeds, but I've never known anyone to grow a pussy willow from seed. They root very easily from semihardwood or hardwood cuttings in sand or any well-drained media with no hormone treatment.

Pussy willows are more a gift to the spirit and a harbinger of spring than an integral part of the landscape. However, I've slipped them into a couple of sunny but inobtrusive backyard spots so that late each winter I can enjoy their gift and heed their message once again.

Pussy willows are shrubs that should be cut back heavily every few years to encourage lots of vigorous new growth, the growth from which the best cutting stems arise. Siting them in a secret, moist corner that is visited only in spring is important because plants are not all that attractive, tend to sucker to the point of being invasive in continuously moist soils, and are subject to a canker disease that may require digging up and discarding the original plant after a few years. Since they're so easy to root, getting a replacement is usually no problem.

The silvery-white pussies are actually the partially opened male flowers. If allowed to develop, the dark brown scales at the base

Figure 97. *Sambucus canadensis*.

of the pussies will drop off, and a furry male catkin covered with golden-yellow pollen will appear. Anyone who has ever picked pussy willows and placed them in a vase of water and then returned a week or so later has seen this pollen dusting the coffee table. To keep pussy willows at their best, put them in the vase but omit the water. They're among the easiest of dried flowers to grow and maintain.

Sambucus canadensis
Elderberry

Habitat: Moist lowlands and ditches from Florida to Nova Scotia. Zones 4–8.

Propagation: Seeds require 2 months' warm stratification followed by at least 3 months' cool stratification. Like the seeds of many natives with a broad geographic range, those of elderberry vary considerably in germination requirements from one end of the plant's range to the other. Softwood cuttings taken in mid-summer and treated with .5 percent IBA powder root readily.

Figure 98. *Sassafras albidum*.

The loose, graceful habit of this large shrub is almost good enough to make it a welcome part of the garden—almost. Unfortunately, the large creamy white flower heads become so heavy with purplish-black fruit in late summer that stems are broken either by the wind or by the weight of birds coming to feast.

If you have a pond or streambank in full sun and wish to attract birds as well as butterflies, do consider elderberries. Cutting the whole bush to the ground every other year will usually keep a clump of elderberry bushes in check. While the raw fruit is too bitter for my palate, I've had good jam and passable wine made from elderberries, and I've also enjoyed some tasty elderflower fritters. In my New England youth, elderberry stems were cut into 6-inch lengths, the pith was forced out of them, and then they were used as "spiles" to direct the sap out of maple trees into buckets during sugaring off time.

Sassafras albidum
Sassafras

Habitat: Fencerows and edges of woodlands from Florida to Maine. Zones 4–8.
Propagation: Seeds require 3–4 months'

cool stratification and then germinate in late spring, seeming to take their time even once the soil has warmed. Good luck beating the birds to the fruit.

Root cuttings taken in late winter while the soil is still cold, stored 3 to 4 weeks at 40–50° in nearly dry sand, and then planted in a well-drained site have proven successful. Patience is required, as sprouts are often slow to show.

Sassafras leaves come in three forms, one that looks like a right-handed mitten, one that looks like a left-handed mitten, and one that has three lobes. Once seen, the shapes are remembered by most people. If seen in fall, the colors, which range from bright, clear yellow through orange and pink to scarlet, are unforgettable. However, my favorite characteristics of this small tree are its yellow-green flower balls that appear in spring just as we are hungering for signs of renewal from the natural world and the fragrance of crushed leaves, stems, or roots.

At one time, sassafras was sought as a cure for most human ailments. Today, the tea brewed from its roots is still considered pleasant and thought of as a spring tonic, but few believe it cures much. A few years ago the government decided sassafras might be hazardous, so root beer now has a different flavoring and the tea has all but disappeared from stores. The concern over potential health hazards seems not to have extended to lovers of Cajun cooking, however, as young leaves are still ground for the filé used in Louisiana gumbo. In areas where black locust and red cedar don't grow, the rot-resistant wood of sassafras has been used for fence posts and trellises. All in all, sassafras is quite a little tree.

Sassafras should be tried in more gardens, especially where the afternoon autumn sun can light up the brilliant foliage. Transplant seedlings or small container-grown plants in early spring. Deep tap roots make transplanting larger trees difficult. Sassafras grows best in well-drained to dry conditions, but will tolerate occasionally moist feet.

Sorbus americana
American mountain ash

Habitat: Moist mountain soils, southern Appalachian balds, Georgia to Newfoundland. Zones 2–6.
Propagation: Seeds require 3 months' warm stratification followed by 2–3 months' cool stratification. Even then, germination is sporadic and may occur over two seasons.

Figure 99. *Sorbus americana*.

The bright red to red-orange berries covering these small trees are responsible for almost as many October photographs along some parts of the Blue Ridge Parkway as is fall foliage. However, throughout most warmer areas these trees are better left in the wild. When brought into cultivation at lower elevations, they are usually very short-lived. They succumb to fireblight, mildew, and borers as well as a number of other diseases and insects, while never quite producing the spectacular fruit displays seen in the wild.

Spiraea tomentosa
Hardhack

Habitat: Wet areas, particularly meadows and the edge of swamps, from the Georgia mountains to Nova Scotia. Zones 4–8.
Propagation: Seeds require no special treatment. Softwood cuttings root readily without hormone.

Although this native spiraea is listed as reaching a height of 6 feet, I've rarely seen it grow much beyond 3 or 4. Its deep pink to rosy-purple flowers complement other midsummer wildflowers when a deep-pink-flowering native shrub may be hard to find.

Choose a moist site in full sun for hardhack to perform at its best. It will tolerate normal garden moisture but dwindles in the shade. Rather than mixing this spiraea with other shrubs and trees, however, try hardhack in a wildflower border. Even purple coneflower can't produce the colorful shades of hardhack. A similar plant, Japanese spiraea, has naturalized throughout the region. While the colors are nearly the same, the Japanese form has a flat-topped flower head (corymb). Hardhack flowers are in a rounded plume (panicle).

Stewartia malacodendron
Silky camellia

Habitat: Along coastal woodland streams and occasionally in rich woods from Florida to Virginia. Zones 7–9.
Propagation: Germination is variable, even when you have good seeds. Warm stratification followed by cool stratification is required for success. Good results have been reported when seeds are given 3–5 months' warm stratification followed by 2–3 months' cool stratification and then watered heavily to wash away natural chemical germination inhibitors. I suggest planting in the fall in a well-marked and protected spot and then waiting two years.

Softwood cuttings are relatively easy to root. However, they should not be moved following rooting until they have gone through normal dormancy and are growing leaves again the following spring. Even then, I have only been able to keep one rooted stewartia cutting alive for as long as two years. But I'm still trying.

Figure 100. *Stewartia malacodendron.*

When horticulturists are lucky enough to see this rare plant flowering in the wild, footsteps and conversation cease. We involuntarily smile (grin?) and inhale to catch our breath. Few of us have had the privilege of viewing wild stewartia, but seeing it in gardens elicits a similar response. I've seen mountain camellia (*Stewartia ovata*) rarely in the wild, and I've never seen silky camellia under those circumstances, but I have enjoyed them both thanks to the efforts of talented southern gardeners.

S. malacodendron is a large shrub or small tree at maturity. It has the glossy green character of its relatives, the true camellias. Unlike camellias, however, stewartias are deciduous. The flowers are 2–3 inches across, white, waxy, and often borne in abundance from mid-summer into early fall on well-cared-for garden specimens. Silky camellia stamens are a startling purple contrasting with the white petals. Mountain camellia is reported to have whitish to purple filaments supporting the stamens, but all that I've seen have been golden.

If you can grow a stewartia, give it a place of honor in your garden, a place where you walk in the summertime. Move plants when young, and only in late winter or earliest spring. To be at its best, silky camellia prefers high shade during the heat of the day but thrives on early morning sunshine. Soil should be the acidic, humusy sort you would choose for your best rhododendrons, with excellent drainage, but it should never be allowed to get dusty dry. You should apply fertilizer sparingly, using less than for rhododendrons. This may seem like a lot of trouble, but stewartias are worth it.

Styrax americana
Snowbell, storax

Habitat: Almost always found bordering coastal plain streams and fresh water wetlands from Florida to Virginia, with one isolated and possibly hardier population reported in Illinois. Zones 6–8?

Propagation: Germination attempts by friends have yielded two forms of success: (1) planted in fall, seedlings emerge the second spring, suggesting a need for warm stratification followed by cool stratification, and (2) fresh seeds planted immediately germinated the next spring. Therefore, I suggest that you plant seeds in the fall. If they don't germinate the first spring, expect them the second. Softwood cuttings root readily when treated with .5–.8 percent IBA powder.

One squintingly bright, frigid March day, I entered a friend's high mountain greenhouse and was immediately enveloped in the delightful fragrance of our native snowbell. The plant hasn't proven hardy where my friend lives, but the spiritual boost given by snowbell—blooming when the weather has been spring for weeks elsewhere but we are often still shoveling snow—has earned it a place in one upper Blue Ridge greenhouse.

Wild snowbells flower in mid-spring, producing a cloud filled with thousands of white reflex-petaled flowers shortly after dogwoods bloom. When competition from other shrubs and trees is kept in check, snowbell will become a large shrub offering a sweet but not overpowering fragrance missing in many spring shrubs. In addition,

Figure 101. *Styrax americana*.

it will grow with its roots in water and offers an early source of nectar to diverse pollinating creatures.

Vaccinium corymbosum
Highbush blueberry

Habitat: Sunny areas with acid soils near water, including mountain balds, edges of bogs, ponds, streams, and swamps, Florida to Nova Scotia. Zones 3–8.

Propagation: Seeds are varyingly reported to need no special treatment and to need stratification! Why worry about it? I think growing any plant as variable as a blueberry from seed is foolish unless you're a plant breeder. If you are going to propagate blueberries, find a bush you want and root cuttings. Softwood cuttings root with .1 percent IBA powder; semihardwood cuttings root easily with .8 percent IBA powder.

Blueberries have one of the greatest potentials in the edible or wildlife-attracting landscape. The fruit attracts both flying and furry wildlife, and the rich to brilliant red fall foliage is better than that of most other deciduous shrubs.

Dr. Sam Jones lists twenty *Vaccinium* species native to the Southeast. Some have been

Figure 102. *Vaccinium corymbosum*.

named and released as effective ground covers, while others have such interesting names as bearberry, buckberry, cranberry, deerberry, farkleberry, and sparkleberry. None, however, is huckleberry. The huckleberries, while closely related and sometimes tasty, are in the genus *Gaylussacia*. The huckleberry with the greatest landscape potential is *G. brachycera*, the box huckleberry. It is a low-growing upland native, excellent as a ground cover under plants preferring acid soils like the tall rhododendrons.

Blueberry culture in the landscape is essentially the same as that for azaleas. Before choosing a blueberry for your landscape, check with your local county agent or nursery for variety names and planting instructions. Plant breeders have done a superb service in producing well-adapted fruitful varieties of these native North American shrubs. There are usually varieties that have proven themselves in each locale. In Zones 8 and 9, rabbiteye blueberries (*V. ashei*) are the preferred choice, while highbush hybrids do better in cooler climates. Getting the best combination of varieties to ensure pollination and top yield is important only if you want fruit as well as the fiery red fall foliage.

Figure 103. *Viburnum alnifolium*.

Viburnum

"A garden without a viburnum is akin to life without music and art," writes Mike Dirr. While this statement may seem extreme, my garden has its viburnums, and I understand Dirr's enthusiasm. The dozen or more species (even botanists haven't agreed which are separate species and which are not) native to the eastern United States range from low shrubs to trees with showy flowers and fruit that can be pink to red or blue to black. There are species that tolerate acid or alkaline soils, moist or dry sites, and sunny or shady locations. What greater versatility can we ask of a genus?

Viburnum alnifolium (lantanoides)

Hobblebush, witch-hobble

Habitat: Moist, shady woodlands and upland roadsides from the Georgia mountains to Maine. Zones 4–7.

Propagation: Seeds of most viburnums are difficult to germinate. They require 3–5 months' warm stratification followed by 2–3 months' cool stratification. Only plant breeders should try seeds since most viburnums root easily from mid-summer semi-hardwood cuttings using .3 percent IBA powder. Seeds aren't worth the trouble when cuttings are so easy.

Figure 104. *Viburnum cassinoides*.

Hobblebush is an open, medium-sized shrub, usually no more than 6 feet tall, but it can reach twice that height. It is best used where its white flowers can be viewed from a distance against woodland shade.

The flower heads have a lace-doily effect similar to that of some hydrangeas and contrast well with the medium green foliage. The fruit is red in late summer, turning to purplish black, but is usually gone by fall. In autumn, foliage ordinarily turns to a deep claret red, which can be effectively contrasted with sassafras or striped maple.

Viburnum cassinoides
Witherod viburnum

Habitat: Moist soils, including bogs and springheads, from Florida to Maine. Zones 4–7.
Propagation: Same as for *V. alnifolium*.

This dense 6-foot shrub blooms white in early summer and will tolerate wetter soil than any other native viburnum. However, its major ornamental characteristics are not necessarily the flowers. When well grown, the fruit is interesting and can verge on spectacular. During the summer the fruit

Figure 105. *Viburnum prunifolium*.

changes from green to an iridescent pink to red; then it becomes blue in late summer or early fall before turning black. Sometimes all these colors are present in the same fruiting cluster. The fall foliage is also variable, turning dull crimson most often, but it may also be orange-red or purple, sometimes with all of these colors on the same bush.

Viburnum prunifolium
Black haw, plum leaf viburnum

Habitat: Woodland edges from Florida to New England. Zones 4–8.

Propagation: Similar to that of *V. alnifolium*, but extend the warm stratification period to about 6 months.

Black haw is a small tree or large shrub with stiff branches and twiggy growth thought by some to resemble hawthorn. Since it has nearly black fruit, the common name became black haw, that is, black-fruited hawthorn.

V. prunifolium must have reasonably well drained soil to grow well but is tolerant of light conditions from full sun to heavy shade. For best flowers and fruit, however, provide at least one-half day of sunlight. The clusters of white flowers appear in late

Figure 106. *Viburnum rafinesquianum*.

spring as leaves are expanding. The fruit is yellowish before turning blue-black in late summer. The black fruit contrasts nicely with red fall foliage if it hasn't all been stripped away by birds or jelly makers.

Viburnum rafinesquianum
Downy arrowwood

Habitat: Dry woodlands from Georgia to Quebec. Zones 3–8.
Propagation: Same as for *V. prunifolium*. Seeds sometimes require 3 years to germinate.

Downy arrowwood and arrowwood, *V. dentatum*, are medium-sized shrubs often used as hedges or screens in the Midwest. The arrowwoods will tolerate a wide range of soils and rugged winter conditions as well as more salt than other viburnums, but not direct ocean spray.

Flowers, blooming slightly after dogwoods, are creamy white contrasted with medium green leaves. Fruit is blue-black, occasionally striking, and attractive to birds. The multitude of straight stems on this plant results in the name arrowwood.

Appendix One

Landscape Fertilizer Rates

| Fertilizer | Amount of Fertilizer to Apply (in Cups)[a] | |
	Maintenance Rate (1 lb. N/1,000 sq. ft.)	Growth Rate (2 lb. N/1,000 sq. ft.)
5-10-10	32 (2 gal.)	64 (4 gal.)
10-10-10	16 (1 gal.)	32 (2 gal.)
Bone meal (4-12-0)	40 (2.5 gal.)	80 (5 gal.)
Blood meal (12-0-0)	20 (1 gal. + 2 cups)	40 (2.5 gal.)
Cottonseed meal (6-1-1)	44 (2.75 gal.)	88 (5.5 gal.)
Composted cow manure (.5-.5-.5)	400 (25 gal.)	800 (50 gal.)
Superphosphate[b]	5 cups 0-20-0 or 2 cups 0-46-0/100 sq. ft. equals 250 lb./acre, the rate often suggested where phosphorous is limiting. Need varies widely throughout the region.	
Dolomitic limestone[b]	5 lb./100 sq. ft. is slightly more than 1 ton/acre, the amount often suggested in soil test reports. 7.25 cups (a half-gallon milk carton 90 percent full) equals 5 lb.	

a. Apply one-third of the amount in mid- to late fall and the other two-thirds in late winter or early spring. Spread the fertilizer uniformly over the area covered by shrub and tree branches, starting 6 in. to 1 ft. from the base of the plant. No fertilizer should touch the base of the plant.

b. Rates for phosphate and lime are given per 100 sq. ft. because these fertilizers are so often applied when adding only a few plants to a border. Both should be mixed thoroughly with the soil before planting.

Appendix Two

Landscape Sizes and Germination Requirements

Plant	Shrub		Tree		Scarify[a]	Stratify[a]	
	Small (< 6 ft.)	Large	Small (< 35 ft.)	Large		Warm	Cool
Acer leucoderme			●				●
Acer pennsylvanicum		●	●				●
Aesculus octandra				●			●
Aesculus parviflora		●					●
Aesculus pavia		●	●				
Amelanchier arborea			●				●
Amelanchier canadensis		●	●				●
Aralia spinosa		●	●				●
Aronia arbutifolia		●					●
Baccharis halimifolia	●	●					
Callicarpa americana	●	●				●	
Calycanthus floridus		●					●
Catalpa bignonioides				●			
Ceanothus americanus	●						●
Cephalanthus occidentalis	●	●					

Plant	Shrub		Tree		Scarify[a]	Stratify[a]	
	Small (< 6 ft.)	Large	Small (< 35 ft.)	Large		Warm	Cool
Cercis canadensis			•		•		•
Chionanthus virginicus		•	•			•	•
Cladrastis lutea				•	•		•
Clethra acuminata		•	•				
Clethra alnifolia	•	•					
Cornus alternifolia			•			•	•
Cornus amomum		•					•
Cornus florida			•				•
Crataegus phaenopyrum			•			•	•
Cyrilla racemiflora		•					
Diervilla sessilifolia	•	•					
Diospyros virginiana			•				•
Euonymus americanus		•				•	•
Fothergilla gardenii	•					•	•
Fothergilla major		•				•	•
Gordonia lasianthus			•				
Halesia carolina			•	•		•	•
Halesia diptera			•			•	•
Hamamelis virginiana		•				•	•
Hydrangea arborescens	•	•					
Hydrangea quercifolia		•					
Hypericum prolificum	•						
Ilex decidua		•	•			•	
Ilex opaca			•	•		•	

Plant	Shrub		Tree		Scarify[a]	Stratify[a]	
	Small (< 6 ft.)	Large	Small (< 35 ft.)	Large		Warm	Cool
Ilex verticillata	•	•				•	
Ilex vomitoria	•	•	•			•	
Itea virginica	•						
Kalmia latifolia	•	•					
Leucothoe fontanesiana	•						
Lindera benzoin		•				•	•
Liriodendron tulipifera				•			•
Magnolia acuminata				•			•
Magnolia grandiflora				•			•
Magnolia virginiana		•	•				•
Malus angustifolia			•				•
Myrica cerifera	•	•	•				•
Myrica pennsylvanica	•	•					•
Ostrya virginiana			•			•	•
Oxydendrum arboreum		•	•				
Philadelphus inodorus		•					
Physocarpus opulifolius		•					
Pieris floribunda	•						
Pinckneya bracteata			•				
Prunus caroliniana		•	•				•
Prunus pennsylvanica			•				•
Prunus serotina			•	•			•
Rhamnus caroliniana			•	•			•
Rhododendron carolinianum	•	•					

Plant	Shrub Small (< 6 ft.)	Shrub Large	Tree Small (< 35 ft.)	Tree Large	Scarify[a]	Stratify[a] Warm	Stratify[a] Cool
Rhododendron catawbiense	●	●					
Rhododendron maximum		●	●				
Rhododendron minus	●	●					
Rhododendron arborescens		●	●				
Rhododendron atlanticum	●						
Rhododendron calendulaceum	●	●					
Rhododendron canescens	●	●					
Rhododendron periclymenoides		●					
Rhododendron roseum	●	●					
Rhododendron vaseyi		●					
Rhododendron viscosum	●	●					
Rhus glabra		●			●		
Robinia hispida		●			●		
Robinia pseudoacacia			●		●		
Rosa carolina	●						●
Rosa virginiana	●						●
Rubus argutus		●					
Rubus odoratus	●	●					
Salix discolor		●					
Sambucus canadensis		●				●	●
Sassafras albidum			●				●
Sorbus americana			●			●	●

Plant	Shrub		Tree		Scarify[a]	Stratify[a]	
	Small (< 6 ft.)	Large	Small (< 35 ft.)	Large		Warm	Cool
Spiraea tomentosa	●						
Stewartia malacodendron		●	●			●	●
Styrax americana		●	●			●	●
Vaccinium corymbosum		●					●
Viburnum alnifolium	●	●				●	●
Viburnum cassinoides	●					●	●
Viburnum prunifolium		●	●			●	●
Viburnum rafinesquianum		●				●	●

a. Neither scarification nor stratification is required if no dot appears in either column.

Appendix Three

Showy Native Shrubs and Trees for Moist Sites

All of the plants in this list will tolerate moist soils. However, some will grow or even thrive with their roots regularly flooded. Check individual plant listings in the text for more information.

Aesculus pavia

Amelanchier canadensis

Aronia arbutifolia

Baccharis halimifolia

Catalpa bignonioides

Cephalanthus occidentalis

Clethra acuminata

Clethra alnifolia

Cornus amomum

Crataegus phaenopyrum

Cyrilla racemiflora

Euonymus americanus

Fothergilla gardenii

Leucothoe fontanesiana

Lindera benzoin

Magnolia grandiflora

Magnolia virginiana

Myrica pennsylvanica

Ostrya virginiana

Physocarpus opulifolius

Pinckneya bracteata

Rhamnus caroliniana

Rhododendron arborescens

Rhododendron atlanticum

Rhododendron vaseyi

Rhododendron viscosum

Rosa virginiana

Rubus odoratus

Salix discolor

Sambucus canadensis

Spiraea tomentosa

Vaccinium corymbosum

Viburnum cassinoides

Appendix Four

Showy Native Shrubs and Trees for Dry Sites

Most of the plants on this list will have to be regularly irrigated for the first year after planting in order to become established and withstand the frequently droughtlike conditions of dry sites.

- Acer leucoderme
- Aralia spinosa
- Ceanothus americanus
- Chionanthus virginicus
- Diervilla sessilifolia
- Diospyros virginiana
- Hypericum prolificum
- Kalmia latifolia
- Myrica cerifera
- Oxydendrum arboreum
- Rhus glabra
- Robinia hispida
- Robinia pseudoacacia
- Rosa carolina
- Rubus argutus
- Sassafras albidum
- Viburnum rafinesquianum

Appendix Five

Showy Native Shrubs and Trees That Attract Wildlife

The plants in this list attract birds, butterflies, or small animals.

Aesculus octandra	*Diospyros virginiana*	*Rhamnus caroliniana*
Aesculus parviflora	*Hydrangea arborescens*	*Rhododendron vaseyi*
Aesculus pavia	*Hypericum prolificum*	*Rosa carolina*
Amelanchier arborea	*Ilex decidua*	*Rosa virginiana*
Amelanchier canadensis	*Ilex opaca*	*Rubus argutus*
Aralia spinosa	*Ilex verticillata*	*Sambucus canadensis*
Aronia arbutifolia	*Ilex vomitoria*	*Sassafras albidum*
Cephalanthus occidentalis	*Itea virginica*	*Sorbus americana*
Chionanthus virginicus	*Lindera benzoin*	*Spiraea tomentosa*
Clethra acuminata	*Magnolia grandiflora*	*Vaccinium corymbosum*
Clethra alnifolia	*Malus angustifolia*	*Viburnum alnifolium*
Cornus amomum	*Prunus caroliniana*	*Viburnum cassinoides*
Cornus florida	*Prunus pennsylvanica*	*Viburnum prunifolium*
Crataegus phaenopyrum	*Prunus serotina*	*Viburnum rafinesquianum*

Appendix Six

Showy Native Shrubs and Trees That Will Tolerate Neutral or Slightly Alkaline Soils

Most of the plants listed here perform best in slightly acid soils, pH 6.5 to 6.8, but will tolerate less acidic soils.

Amelanchier arborea	*Hydrangea quercifolia*	*Physocarpus opulifolius*
Aronia arbutifolia	*Hypericum prolificum*	*Prunus caroliniana*
Baccharis halimifolia	*Ilex vomitoria*	*Rhus glabra*
Cercis canadensis	*Lindera benzoin*	*Robinia hispida*
Cladrastis lutea	*Magnolia grandiflora*	*Robinia pseudoacacia*
Clethra alnifolia	*Malus angustifolia*	*Rosa carolina*
Crataegus phaenopyrum	*Myrica cerifera*	*Rosa virginiana*
Euonymus americanus	*Myrica pennsylvanica*	*Rubus argutus*
Hydrangea arborescens	*Philadelphus inodorus*	*Salix discolor*

Appendix Seven

Sources of Rooting Hormones and Other Horticultural Supplies

Cassco
P.O. Box 3508
Montgomery, AL 36193

Coor Farm Supply
P.O. Box 525
Smithfield, NC 27577

Forestry Suppliers, Inc.
P.O. Box 8397
Jackson, MS 39284-8397

E. C. Geiger Co.
Route 63, Box 285
Harleysville, PA 19438

Good Prod Sales, Inc.
825 Fairfield Avenue
Kenilworth, NJ 07033

Great Western Bag Co.
Route 10, Box 206B
McMinnville, TN 37110

Horticultural Products
P.O. Box 35038
1307 W. Morehead Street
Charlotte, NC 28235-5038

A. H. Hummert
2746 Chauteau Avenue
St. Louis, MO 63103

Park Seed Co.
Cokesbury Road
Greenwood, SC 29647

VJ Growers Supply
Box 240925
4941 Chastain Avenue
Charlotte, NC 28224

Appendix Eight

Sources of Seeds for Native Woody Plants

W. Atlee Burpee & Co.
300 Park Avenue
Warminster, PA 18974

Herbst Seeds
307 Number 9 Road
Fletcher, NC 28732

Park Seed Co.
Cokesbury Road
Greenwood, SC 29647

F. W. Schumacher Co.
Horticulturists
Sandwich, MA 02543

Sheffield's Seed Co.
273 Route 34
Locke, NY 13092

Appendix Nine

Nursery Sources of Native Woody Plants

Carlson's Gardens
Box 305
South Salem, NY 10590

Forestfarm
990 Tetherow Road
Williams, OR 97544

Holbrook Farm
P.O. Box 368
Fletcher, NC 28732

Lamtree Farm
Route 1, Box 162
Warrensville, NC 28693

Magnolia Nursery
Route 1, Box 87
Chunchula, AL 36521

Native Gardens
Route 1, Box 494
Greenback, TN 37742

Native Nurseries
1661 Centerville Road
Tallahassee, FL 32308

Niche Gardens
1111 Dawson Road
Chapel Hill, NC 27516

Roslyn Nursery
Box 69
Roslyn, NY 11576

Salter Tree Farm
Route 3, Box 1332
Madison, FL 32340

Transplant Nursery
Parkertown Road
Lavonia, GA 30553

Wayside Gardens
1 Garden Lane
Hodges, SC 29695-0001

Woodlanders, Inc.
1128 Colleton Avenue
Aiken, SC 29801

Selected References

Borland, Hal, and Les Line. *A Countryman's Flowers.* New York: Alfred A. Knopf, 1981.

Borland, Hal, and Les Line. *A Countryman's Woods.* New York: Alfred A. Knopf, 1983.

Brown, Claud L., and L. Katherine Kirkman. *Trees of Georgia and Adjacent States.* Portland, Ore.: Timber Press, 1990.

Browse, Philip Macmillan. *Plant Propagation.* New York: Simon & Schuster, 1988.

Dirr, Michael A. *Manual of Woody Landscape Plants.* Champaign, Ill.: Stipes Publishing Co., 1990.

Dirr, Michael A., and Charles W. Heuser, Jr. *The Reference Manual of Woody Plant Propagation.* Athens, Ga.: Varsity Press, 1989.

Duncan, Wilbur H., and Marion B. Duncan. *Trees of the Southeastern United States.* Athens, Ga.: University of Georgia Press, 1988.

Flint, Harrison. *The Country Journal Book of Hardy Trees.* Brattleboro, Vt.: Country Journal Publishing Co., 1983.

Foote, Leonard E., and Samuel B. Jones. *Native Shrubs and Woody Vines of the Southeast.* Portland, Ore.: Timber Press, 1989.

Frederick, William H., Jr. *100 Great Garden Plants.* Portland, Ore.: Timber Press, 1975.

Hartmann, Hudson, Dale Kester, and Fred Davies, Jr. *Plant Propagation Principles and Practices.* Englewood Cliffs, N.J.: Prentice Hall, 1990.

Hillier's Manual of Trees & Shrubs. Newton Abbot, Devon, Eng.: David & Charles, 1974.

Jaynes, Richard A. *Kalmia: The Laurel Book II.* Portland, Ore.: Timber Press, 1988.

Justice, William S., and C. Ritchie Bell. *Wild Flowers of North Carolina.* Chapel Hill: University of North Carolina Press, 1968.

Radford, Albert E., Harry E. Ahles, and C. Ritchie Bell. *Manual of the Vascular Flora of the Carolinas.* Chapel Hill: University of North Carolina Press, 1968.

Smith, Richard M. *Wild Plants of America.* New York: John Wiley & Sons, 1989.

Index of Scientific Names

Acer
   leucoderme, 71, 163, 171
   pennsylvanicum, 72, 163
Aesculus
   octandra, 73, 163, 173
   parviflora, 74, 163, 173
   pavia, 75, 163, 169, 173
Agarista populifolia, 107
Amelanchier
   arborea var. *arborea*, 75, 163, 173, 175
   arborea var. *laevis*, 75, 163, 173, 175
   canadensis, 75, 163, 169, 173
   grandiflora, 76, 77
   obovalis, 77
Aralia spinosa, 77, 163, 171, 173
Aronia arbutifolia, 79, 163, 169, 175

Baccharis halimifolia, 79, 163, 169, 175

Callicarpa
   americana, 80, 163
   dichotoma, 60
Calycanthus floridus, 81, 163
Catalpa bignonioides, 82, 163, 169
Ceanothus americanus, 82, 163, 171

Cephalanthus occidentalis, 83, 163, 169, 173
Cercis canadensis, 3, 4, 84, 164, 175
Chionanthus virginicus, 16, 29, 85, 164, 171, 173
Cladrastis lutea (kentuckea), 85, 164, 175
Clethra
   acuminata, 86, 164, 169, 173
   alnifolia, 87, 164, 169, 173
Cornus
   alternifolia, 88, 164
   amomum, 47, 88, 164, 169, 173
   florida, 18, 47, 89, 164, 173
Crataegus
   phaenopyrum, 91, 164, 169, 173, 175
   viridis, 91
Cyrilla racemiflora, 91, 164, 169

Diervilla sessilifolia, 92, 164, 171
Diospyros virginiana, 92, 164, 171, 173

Euonymus americanus, 94, 164, 169, 175

Fothergilla
   gardenii, 94, 164, 169
   major, 95, 164

Gaylussacia brachycera, 156
Gordonia lasianthus, 95, 164

Halesia
   carolina (tetraptera), 29, 30, 96, 164
   diptera, 97, 164
Hamamelis virginiana, 19, 98, 164
Hydrangea
   arborescens, 99, 164, 173, 175
   quercifolia, 99, 164, 175
Hypericum prolificum, 100, 164, 171, 173, 175

Ilex
   coriacea, 103
   decidua, 101, 164, 173
   glabra, 103
   opaca, 102, 164, 173
   serrata, 105
   verticillata, 102, 164, 173
   vomitoria, 102, 165, 173, 175
Itea virginica, 105, 165, 173

Kalmia latifolia, 106, 165, 171

Leucothoe
   axillaris, 107
   fontanesiana (catesbaei), 107, 165, 169
   populifolia, 107

Lindera benzoin, 108, 165, 169, 173, 175
Liriodendron tulipifera, 109, 165

Magnolia
  acuminata, 111, 165
  ashei, 111
  fraseri, 111
  grandiflora, 61, 110, 112, 165, 169, 173, 175
  macrophylla, 111
  pyramidata, 111
  tripetala, 111, 112
  virginiana, 47, 110, 114, 165, 169
Malus angustifolia, 114, 165, 173, 175
Myrica
  cerifera, 115, 165, 171, 175
  pennsylvanica, 116, 165, 169, 175

Ostrya virginiana, 117, 165, 169
Oxydendrum arboreum, 118, 165, 171

Philadelphus
  coronarius, 119
  inodorus, 119, 165, 175
  lemoinei, 119
  nivalis, 119
  virginalis, 119
Physocarpus opulifolius, 119, 165, 169, 175
Pieris floribunda, 121, 165
Pinckneya bracteata (pubens), 122, 165, 169

Prunus
  americana, 124
  angustifolia, 124
  avium, 124
  caroliniana, 124, 165, 173, 175
  pennsylvanica, 125, 165, 173
  serotina, 125, 165, 173

Rhamnus caroliniana, 126, 165, 169, 173
Rhododendron
  arborescens, 133, 134, 166, 169
  atlanticum, 133, 135, 166, 169
  bakeri, 133
  calendulaceum, 29, 133, 136, 166
  canescens, 133, 137, 166
  carolinianum, 128, 132, 165
  catawbiense, 129, 166
  maximum, 39, 131, 166
  minus, 128, 132, 166
  periclymenoides (nudiflorum), 132, 138, 166
  roseum, 133, 140, 166
  vaseyi, 9, 11, 133, 140, 166, 169, 173
  viscosum, 133, 141, 166, 169
Rhus
  glabra, 142, 166, 171, 175
  typhinia, 142
Robinia
  hispida, 143, 166, 171, 175
  pseudoacacia, 3, 144, 166, 171, 175
Rosa
  carolina, 145, 166, 171, 173, 175
  palustris, 145
  rugosa, 145
  virginiana, 146, 166, 169, 173, 175
Rubus
  argutus, 147, 166, 171, 173, 175
  odoratus, 148, 166, 169

Salix discolor, 31, 149, 166, 169, 175
Sambucus canadensis, 150, 166, 169, 173
Sassafras albidum, 151, 166, 171, 173
Sorbus americana, 152, 166, 173
Spiraea tomentosa, 153, 167, 169, 173
Stewartia
  malacodendron, 153, 167
  ovata, 154
Styrax americana, 154, 167

Tsuga
  canadensis, 61
  carolina, 61

Vaccinium
  ashei, 156
  corymbosum, 155, 167, 169, 173
Viburnum
  alnifolium (lantanoides), 157, 167, 169, 173
  cassinoides, 158, 167, 169, 173
  dentatum, 160
  prunifolium, 159, 167, 173
  rafinesquianum, 160, 167, 171, 173

Index of Common Names

Alder, black, 102
Allspice, Carolina, 81
American snowdrop tree, 97
'Annabelle' (*Hydrangea arborescens*), 100
'Arnot' (*Robinia hispida*), 143
Arrowwood, downy, 160
'Athens' (*Calycanthus floridus*), 81
'Atrosanguinea' (*Aesculus pavia*), 75
'Autumn Brilliance' (*Amelanchier grandiflora*), 77
'Autumn Glow' (Ilex verticillata), 105
Azalea
  Baker's, 133
  clove, 140
  coastal, 135
  Cumberland, 133
  dwarf, 135
  flame, 29, 136
  Florida pinxter, 137
  hoary, 137
  piedmont, 137
  pinkshell, 9, 11, 140
  pinxterbloom, 138
  roseshell, 140
  smooth, 134
  swamp, 141
  sweet, 134

Bayberry, 116
Bearberry, 156
Beautyberry, 80
Black haw, 159
Black walnut, 59
Blackberry, highbush, 147
'Blue Mist' (*Fothergilla gardenii*), 95
Blueberry
  highbush, 155
  rabbiteye, 156
'Brilliantissima' (*Aronia arbutifolia*), 79
'Brouwer's Beauty' (*Pieris* hybrid), 122
Buckberry, 156
Buckeye
  bottlebrush, 74
  red, 75
  yellow, 73
Buckthorn, Carolina, 126
Bull bay, 112
Buttonbush, 83

Calico bush, 106
'Carousel' (*Kalmia latifolia*), 107
Catalpa, southern, 82
Chalk maple, 71
'Cherokee Chief' (*Cornus florida*), 90
'Cherokee Princess' (*Cornus florida*), 90

Cherry
  black, 125
  choke, 124
  fire, 125
  Indian, 126
  pin, 125
Cherry laurel, Carolina, 124
Chickasaw plum, 124
Chokeberry, red, 79
Cinnamonbark clethra, 86
Crabapple, southern, 114
Cranberry, 156
Cucumber tree, 111

Deerberry, 156
'Delaware Blue' (*Rhododendron viscosum*), 141
Devil's walking stick, 77
Doghobble, 107
Dogwood
  flowering, 18, 47, 89
  pagoda, 88
  silky, 47, 88

Elder, salt marsh, 79
Elderberry, 150
Election pink, 140
'Elf' (*Kalmia latifolia*), 107

Farkleberry, 156
Fetterbush, 121
Fever tree, 122

Fothergilla
   dwarf, 94
   large, 95
Fringe tree, 16, 85

Gallberry, 103
'Girard's Rainbow' (*Leucothoe fontanesiana*), 106
Gopherwood, 85
Grancy Greybeard, 85
'Grey's Weeping' (*Ilex vomitoria*), 104
Groundsel tree, 79

Hardhack, 153
'Harmony' (*Hydrangea quercifolia*), 100
Hawthorn
   green, 91
   Washington, 91
Hearts a bustin', 94
Hemlock
   Carolina, 61, 62
   eastern (Canadian), 61, 62
'Henry's Garnet' (*Itea virginica*), 60, 106
Hercules club, 77
Hobblebush, 157
Holly
   American, 102
   yaupon, 102
Honeysuckle
   bush, 92
   mountain, 92
   swamp, 141
   wild, 138
   yellow, 136
Hornbeam, hop, 117
Huckleberry, 156
'Hummingbird' (*Clethra alnifolia*), 88
Hydrangea
   oak-leaf, 99
   smooth, 99
   wild, 99

Indian bean, 82
Ironwood, 117

Judas tree, 84
June punctatum, 132
Juneberry, 75, 76

'Laciniata' (*Rhus glabra; R. typhinia*), 143
Laurel
   Carolina cherry, 124
   deer-tongue, 128
   great, 131
   mountain, 106
   purple, 129
   red, 129
   white, 131
Leatherwood, 91
Leucothoe
   drooping, 107
   Florida, 107
Lily of the valley tree, 118
Locust
   black, 3, 144
   bristly, 143
   white, 144
   yellow, 144

'Magniflora' (*Halesia diptera*), 97
Magnolia
   Ashe, 111
   bigleaf, 111
   Fraser's, 111
   pyramid, 111
   southern, 61, 112
   tulip, 109
   umbrella, 111
Maple
   chalk, 71
   goosefoot, 72
   southern sugar, 71
   striped, 72
Max, 131
Minus, 132
Mock orange, 119

Moosewood, 72
'Mount Airy' (*Fothergilla* hybrid), 95
Mountain andromeda, 121
Mountain ash, American, 152
Mountain camellia, 154
Mountain fetterbush, 121
Mountain laurel, 106
Mountain pepperbush, 86
Mountain snowball, 82
Mountain stewartia, 154
Mulberry, French, 80

New Jersey tea, 82
Ninebark, 119
'Nipmuck' (*Kalmia latifolia*), 107

Old man's beard, 85
'Olympic Fire' (*Kalmia latifolia*), 107

Pagoda dogwood, 88
'Paniculata' (*Clethra alnifolia*), 88
'Pendula' (*Ilex vomitoria*), 104
Pepperbush
   mountain, 86
   sweet, 87
Persimmon, 93
'Pinkspire' (*Clethra alnifolia*), 88
Pinxterbloom, 138
Plum
   Chickasaw, 124
   wild, 124
'Pocahontas' (*Ilex decidua*), 104
Poinsettia tree, 122
Poplar
   tulip, 109
   yellow, 109
Possum haw, 101
'Prince Charles' (*Amelanchier arborea var. laevis*), 75
'Prince William' (*Amelanchier canadensis*), 77
'Pristine' (*Kalmia latifolia*), 107
Punctatum, 128
Pussy willow, 31, 149

Raspberry, flowering, 148
Redbud, 3, 4, 84
Redroot, 82
Rhododendron
   Carolina, 128
   Catawba, 129
Rose
   Carolina, 145
   Cherokee, 145
   dog, 145
   multiflora, 145
   pasture, 145
   Virginia, 146
   wild, 145
'Rosea' (*Clethra alnifolia*), 88
Rosebay
   common, 131
   mountain, 129
'Rubra' (*Cornus florida*), 90

Saint-John's-wort, 100
'Sarah' (*Kalmia latifolia*), 107
Sarvice, 75, 76
Sassafras, 151
'Scarletta' (*Leucothoe fontanesiana*), 107
'Schelling's' (*Ilex vomitoria*), 104

Serviceberry, 75, 76
Sevenbark, 99
Shadblow, 75
Shadbush, 76
Silky-cornel, 88
Silverbell, Carolina, 3, 29, 30, 96
'Snow Queen' (*Hydrangea quercifolia*), 100
Snowbell, 154
'Snowflake' (*Hydrangea quercifolia*), 100
Sorrell tree, 118
Sourwood, 3, 118
Sparkleberry, 105, 156
Spice bush, 108
Storax, 154
Strawberry bush, 94
Strawberry shrub, 81
'Stroke's Dwarf' (*Ilex vomitoria*), 104
Sumac
   smooth, 142
   staghorn, 142
Summersweet, 87
Sweet bay, 47, 114
Sweet pepperbush, 87
Sweet shrub, 81

Titi, 91
Tulip tree, 109

Vergilia, 85
Viburnum
   plum leaf, 159
   witherod, 158
Virginia sweet spire, 105
Virginia willow, 105

'Warren's Red' (*Ilex decidua*), 104
Washington hawthorn, 91
Wax myrtle, southern, 115
Whistlewood, 72
'Winter King' (*Crataegus phaenopyrum*), 91
'Winter Red' (*Ilex verticillata*), 105
Winterberry, 102
Witch hazel, 19, 98
Witch-hobble, 157

Yaupon holly, 102
Yellowwood, 85

General Index

Acid scarification, 15–16

Bottom heat, 37

Climatic factors, 3
Conservation through propagation, 11–12
Cooler rooter, 37–38
Cuttings
    root. See Propagation: root cuttings
    stem. See Propagation: stem cuttings

Dead-heading, 67

Essential elements/nutrients, 49–52

Fermentation, 19
Fertilization
    calculating application rates, 64–66
    of cuttings, 42–43
    in landscape, 49–52, 63–66, 161
    of seedbeds, 25
    of seedlings, 22
    timing of, 64
Flowering, promoting, 67

Grass, 62–63

Hardiness zones, 9–10
Hardwood bark, 48
Hormones, rooting, 32–33

IBA (indolebutyric acid), 33
Identification of plants, 2–3
Irrigation, 24, 54

Limbing up, 46, 66
Limestone, 21, 51

Macronutrients, 49–51
Major elements, 49–51
Media, potting, 21, 26, 31–32
Micronutrients, 52
Minor elements, 52
Mulch, 28, 48, 54, 56–59

NAA (naphthaleneacetic acid), 33
Nitrogen, 49–50
Nutrients, 49–52

Peat, 16, 31, 48
Persimmon pudding, recipe for, 93
Pine bark, 23, 47
Phosphate, 49
Plant communities, 45–46
Plant competition, 59–63
Planting, landscape, 53–59
    backfilling in, 56, 58
    of balled and burlapped plants, 55
    of container-grown plants, 55
    depth of, 54
    hole for, 53
    of rhododendrons, 54
    of rootbound plants, 56–57
    site preparation in, 47
    site selection in, 46
    spacing of, 59
Polarity, maintaining with root cuttings, 41
Potting media (compost), 21, 26, 31–32
Propagation
    root cuttings, 41–42
    sanitation in, 32, 35
    seed, 14–24
        bed preparation, 23
        collection, 18–20
        germination, 24
        processing, 19–20
        scarification, 14–16
        sowing, 20, 23
        storage, 20
        stratification, 16–18
    stem cuttings, 30–39
        care, 37, 39
        gathering, 34
        greenwood, 33
        hardwood, 34
        hormones, 32–33
        media, 31–32
        misting, 37
        preparing, 34–36
        semihardwood, 33
        softwood, 33
        sticking, 36
        timing, 39
        wounding, 35
Pruning
    reasons for, 66–67
    of rhododendrons, 128, 134

Raised beds, 47
Root cuttings. *See* Propagation: root cuttings
Rooting hormones, 32–33

Scarification, 14–16
Secateurs, 35
Secondary elements, 51–52
Secondary nutrients, 51–52
Seeds. *See* Propagation: seed
Site analysis, 46
Soil
    amendments, 23, 47–49
    chemistry, 49–52
    drainage, 47
    pH, 51
    physics, 47–49
    structure, 47–49
    testing, 23
Stem cuttings. *See* Propagation: stem cuttings
Stratification, 16–18
Sulfur, 51
Sumac drink, recipe for, 143

Temperature variation, 3
Transplanting
    of cuttings, 44
    into landscape, 52–56
    of seedlings, 25–26

Vermiculite, 32

Weeds
    in media and mulch, 21, 28, 59
    and plant competition, 62–63
Winter protection
    of container-grown plants, 27
    of garden-grown plants, 28
Wood ashes, 66